ENERGY 2025

The Fifteenth International Conference on Smart Grids, Green Communications and IT Energy-aware Technologies

March 9th –13th, 2025

Lisbon, Portugal

ENERGY 2025 Editors

Vivian Sultan, California State University, Los Angeles, USA

Printed from e-media with permission by:

Curran Associates, Inc.
57 Morehouse Lane
Red Hook, NY 12571

Some format issues inherent in the e-media version may also appear in this print version.

International Academy, Research, and Industry Association (IARIA)
412 Derby Way
Wilmington, DE 19810

Phone: (408) 893-6407
Fax: (408) 527-6351

petre@iaria.org

Additional copies of this publication are available from:

Curran Associates, Inc.
57 Morehouse Lane
Red Hook, NY 12571 USA
Phone: 845-758-0400
Fax: 845-758-2633
Email: curran@proceedings.com
Web: www.proceedings.com

Table of Contents

Impact of Reforestation on Wildfire Severity in California: A GIS-Based Analysis 1
Vivian Sultan and Tyler Kom

Power Grid Vulnerability and Secure Infrastructure for Energy Transmission 7
Vivian Sultan, Philip Caban, Samuel Wong, Aaditya Kandel, and Juan Talome

Analyzing Electric Vehicle Charging Infrastructure Accessibility in Los Angeles, California 11
Vivian Sultan and Oscar Aguillon

Potential for Wind Energy on the Fort Bidwell Reservation 16
Vivian Sultan and Hector Tarango

Security and Attacks on Federated Energy Forecasting 23
Jonas Sievers, Krupali Kumbhani, Thomas Blank, Frank Simon, and Andreas Mauthe

SECNIR: A Multi-Year Electricity Consumption Dataset of 881 French Companies in Islands and Overseas Regions 29
Pierre-Alexis Chevreuil, Antoine Tavant, Pierre Costini, Cedric Simi, and Joris Costes

Disaggregation of Heating and Cooling Energy Consumption via Maximum a Posteriori Estimation 35
Antoine Tavant, Cedric Simi, Pierre-Alexis Chevreuil, Pierre Costini, and Joris Costes

Analysis of Data from a 5G-Based Photovoltaic Plant Monitoring System 41
Sofia Diniz Carvalho, Paulo Cesar Marques Carvalho, Luis M. Fernandez Ramirez, and Paulo Renato X Silva

On Zero-shot Learning in Neural State Estimation of Power Distribution Systems 47
Aleksandr Berezin, Stephan Balduin, Eric MSP Veith, Thomas Oberliessen, and Sebastian Peter

Federated Learning for Distributed Load Forecasting: Addressing Data Imbalance in Smart Grids 53
Alexander Wallis, Sascha Hauke, Hannah Jorg, and Konstantin Ziegler

Offline Reinforcement Learning Agents for Adaptive Reactive Power Control with Renewable Energy Sources 59
Tejashri Bhatt, Stephan Balduin, and Eric MSP Veith

A Systems Dynamics Analysis of Communication Technology Integration in Complex Transactive Energy Systems 68
Atefeh Zarei, Evans Honu, Mo Mansouri, and Philip Odonkor

ENERGY 2025

Foreword

The Fifteenth International Conference on Smart Grids, Green Communications and IT Energy-aware Technologies (ENERGY 2025), held between March 9 - 13, 2025, continued the event considering Green approaches for Smart Grids and IT-aware technologies. It addressed fundamentals, technologies, hardware and software needed support, and applications and challenges.

There is a perceived need for a fundamental transformation in IP communications, energy-aware technologies and the way all energy sources are integrated. This is accelerated by the complexity of smart devices, the need for special interfaces for an easy and remote access, and the new achievements in energy production. Smart Grid technologies promote ways to enhance efficiency and reliability of the electric grid, while addressing increasing demand and incorporating more renewable and distributed electricity generation. The adoption of data centers, penetration of new energy resources, large dissemination of smart sensing and control devices, including smart home, and new vehicular energy approaches demand a new position for distributed communications, energy storage, and integration of various sources of energy.

We take here the opportunity to warmly thank all the members of the ENERGY 2025 Technical Program Committee, as well as the numerous reviewers. The creation of such a high quality conference program would not have been possible without their involvement. We also kindly thank all the authors who dedicated much of their time and efforts to contribute to ENERGY 2025. We truly believe that, thanks to all these efforts, the final conference program consisted of top quality contributions.

Also, this event could not have been a reality without the support of many individuals, organizations, and sponsors. We are grateful to the members of the ENERGY 2025 organizing committee for their help in handling the logistics and for their work to make this professional meeting a success.

We hope that ENERGY 2025 was a successful international forum for the exchange of ideas and results between academia and industry and for the promotion of progress in the fields of smart grids, green communications and IT energy-aware technologies.

We are convinced that the participants found the event useful and communications very open. We also hope that Nice provided a pleasant environment during the conference and everyone saved some time for exploring this beautiful city.

ENERGY 2025 Chairs:

ENERGY 2025 Steering Committee

Eric MSP Veith, Carl von Ossietzky University – Oldenburg, Germany
Dragan Obradovic, Siemens - Corporate Technology, Munich, Germany
Mark Apperley, University of Waikato, New Zealand
Michael Negnevitsky, University of Tasmania, Australia
Vivian Sultan, California State University Los Angeles, USA
Steffen Fries, Siemens, Germany
Philip Odonkor, Stevens Institute of Technology, USA

ENERGY 2025 Publicity Chairs

Francisco Javier Díaz Blasco, Universitat Politècnica de València, Spain
Ali Ahmad, Universitat Politècnica de València, Spain

Sandra Viciano Tudela, Universitat Politecnica de Valencia, Spain
José Miguel Jiménez, Universitat Politecnica de Valencia, Spain

ENERGY 2025

Committee

ENERGY 2025 Steering Committee
Eric MSP Veith, OFFIS e.V. – Oldenburg, Germany
Dragan Obradovic, Siemens - Corporate Technology, Munich, Germany
Mark Apperley, University of Waikato, New Zealand
Michael Negnevitsky, University of Tasmania, Australia
Vivian Sultan, California State University Los Angeles, USA
Steffen Fries, Siemens, Germany
Philip Odonkor, Stevens Institute of Technology, USA

ENERGY 2025 Publicity Chairs
Francisco Javier Díaz Blasco, Universitat Politècnica de València, Spain
Ali Ahmad, Universitat Politècnica de València, Spain
Sandra Viciano Tudela, Universitat Politecnica de Valencia, Spain
José Miguel Jiménez, Universitat Politecnica de Valencia, Spain

ENERGY 2025 Technical Program Committee

Roozbeh Abolpour, Shiraz University, Iran
Kodjo Agbossou, Université du Québec à Trois-Rivières, Canada
Awadelrahman M. A. Ahmed, University of Oslo, Norway
Miltos Alamaniotis, University of Texas at San Antonio, USA
Kamal Al-Haddad, École de technologie supérieure, Montreal, Canada
Ahmed Al-Salaymeh, The University of Jordan, Amman, Jordan
Amjad Anvari-Moghaddam, AAU Energy, Denmark
Mark Apperley, University of Waikato, New Zealand
Parimal Acharjee, NIT Durgapur, India
Paranietharan Arunagirinathan, Clemson University, USA
Ashrant Aryal, Texas A&M University, USA
Adela Bara, Bucharest University of Economic Studies, Department of Economic Informatics and Cybernetics, Bucharest, Romania
Rico Berner, Ambrosys GmbH, Potsdam, Germany
Lasse Berntzen, University of South-Eastern Norway, Norway
Vito Calderaro, University of Salerno, Italy
Hasan Basri Celebi, Hitachi Energy, Sweden
M. Girish Chandra, TCS Research & Innovation, India
Fathia Chekired, CDER-UDES, Algeria
Dana-Alexandra Ciupageanu, National University of Science and Technology POLITEHNICA Bucharest, Romania
Daniele Codetta, University of Piemonte Orientale, Italy
Luigi Costanzo, Università degli Studi della Campania Luigi Vanvitelli, Italy
Fabio D'Agostino, University of Genova, Italy

Thusitha Dayaratne, Monash University, Australia
Payman Dehghanian, The George Washington University, USA
Margot Deruyck, Universiteit Gent - IMEC - WAVES, Belgium
Giovanna Dondossola, RSE, Italy
Virgil Dumbrava, National University of Science and Technology POLITEHNICA Bucharest, Romania
Kevin Ellett, Indiana University, USA
Emin Taner Elmas, IGDIR University, Turkey
Tatiana Endrjukaite, Transport and Telecommunication Institute, Riga, Latvia
Meisam Farrokhifar, Eindhoven University of Technology, The Netherlands
Sébastien Faye, Luxembourg Institute of Science and Technology (LIST), Luxembourg
Wendy Flores-Fuentes, Autonomous University of Baja California, Mexicali, Mexico
Mahmoud Fotuhi-Firuzabad, Sharif University of Technology, Tehran, Iran
Steffen Fries, Siemens, Germany
Vincenzo Galdi, University of Salerno, Italy
Francisco M. Gonzalez-Longatt, University of South-Eastern Norway, Norway
Saman K. Halgamuge, The University of Melbourne, Australia
Yunzhi Huang, Pacific Northwest National Laboratory - U.S. Department of Energy, USA
Md. Minarul Islam, University of Dhaka, Bangladesh
Michael Kuhn, Otto von Guericke University Magdeburg, Germany
Rajat Kumar, Dhirubhai Ambani Institute of information and communication technology, Gandhinagar, India
Tobias Küster, Technische Universität Berlin (TU Berlin), Germany
Sebastian Lawrenz, Clausthal University of Technology, Germany
Duc Van Le, Nanyang Technological University, Singapore
Gerard Ledwich, Queensland University of Technology, Australia
Yiu-Wing Leung, Hong Kong Baptist University, Kowloon Tong, Hong Kong
Qinghua Li, University of Arkansas, USA
Zhenhua Liu, Stony Brook University (SUNY at Stony Brook), USA
Pascal Maussion, Toulouse INP / CNRS LAPLACE, France
HoSoon Min, INTI International University, Malaysia
Hugo Morais, Universidade de Lisboa, Portugal
Fabio Mottola, University of Naples Federico II, Italy
Emmanuel Mudaheranwa, Cardiff University, UK / Rwanda Polytechnic, Rwanda
Gero Mühl, Universitaet Rostock, Germany
Hamidreza Nazaripouya, University of California, Riverside, USA
Michael Negnevitsky, University of Tasmania, Australia
Philip Odonkor, Stevens Institute of Technology, USA
Simona Olmi, Consiglio Nazionale delle Ricerche - Istituto dei Sistemi Complessi, Italy
Claudiu Oprea, Technical University of Cluj-Napoca, Romania
Youssef Ounejjar, ETS, Montreal, Canada
Shalini Pandey, University of Minnesota, USA
Sanjeev Pannala, Washington State University, USA
Pat Pannuto, University of California, San Diego, USA
Thanasis Papaioannou, Athens University of Economics and Business (AUEB), Greece
Marco Pasetti, University of Brescia, Italy
Nilavra Pathak, University of Maryland Baltimore County, USA
Lawrence Pileggi, Carnegie Mellon University, USA
Anand Krishnan Prakash, Lawrence Berkeley National Laboratory, USA

Marco Pruckner, Friedrich-Alexander-University Erlangen-Nürnberg, Germany
Venkata Ramakrishna P., Tata Consultancy Services, India
Anuradha Ravi, University of Maryland Baltimore County, USA
Djamila Rekioua, University of Bejaia, Algeria
Huamin Ren, Kristiania University College, Oslo, Norway
Jan Richling, South Westphalia University of Applied Sciences, Germany
Stefano Rinaldi, University of Brescia, Italy
Carsten Rudolph, Monash University, Australia
Angela Russo, Politecnico di Torino, Italy
Eckehard Schöll, Technische Universität Berlin | Institut für Theoretische Physik, Germany
Tomonobu Senjyu, University of the Ryukyus, Japan
S. Senthilraja, SRM Institute of Science and Technology, India
Farhad Shahnia, Murdoch University, Australia
Hussain Shareef, United Arab Emirates University, UAE
Bhim Singh, Indian Institute of Technology Delhi, India
Vijay Sood, Ontario Tech University, Canada
Vivian Sultan, California State University, Los Angeles, USA
Hongbo Sun, Mitsubishi Electric Research Laboratories, USA
Masoud Taghavi, Technical and Vocational University (TVU), Faculty of Noshahr, Iran
Mehrdad Tahmasebi, Islamic Azad University - Ilam Branch, Iran
Saeed Teimourzadeh, EPRA - Engineering Procurement Research Analysis, Ankara, Turkey
Philipp Thies, University of Exeter, UK
Mihai Tirsu, Institute of Power Engineering, Moldova
Tek Tjing Lie, Auckland University Of Technology, New Zealand
Santiago Torres Contreras, Universidad de Cuenca, Ecuador
Graham Town, Macquarie University, Australia
Quoc Tuan Tran, Paris Saclay University / CEA / INES, France
Navid Vafamand, Shiraz University, Iran
François Vallee, University of Mons, Belgium
Eric MSP Veith, Carl von Ossietzky University, Oldenburg, Germany
Alekhya Velagapudi, University of Pittsburgh's School of Computing and Information, USA
Alexander Wallis, University of Applied Sciences Landshut, Germany
Daniel Wilson, Boston University, USA
Jian Xu, Texas Reliability Entity (Texas RE), USA
Weiwei Yang, Microsoft Research, USA
Sean Yaw, Montana State University, USA

Copyright Information

Impact of Reforestation on Wildfire Severity in California: A GIS-Based Analysis

Vivian Sultan
California State University
Los Angeles, CA USA
email: vsultan3@calstatela.edu

Tyler Kom
California State University
Los Angeles, CA USA
email: tkom@calstatela.edu

Abstract—**California, with its diverse landscapes and increasing wildfire threats, faces a critical challenge in managing the spatial dynamics of tree-planting initiatives and fire risk. This study investigates the spatial relationship between urban tree-planting areas and wildfire-severity zones in California using advanced spatial analysis techniques. By integrating data on fire-hazard severity from CAL FIRE and urban tree-canopy cover from the U.S. Department of Agriculture, the analysis employs the Global Moran's I tool using Inverse Distance, K-Nearest Neighbors, and Fixed Distance Band methods to assess spatial autocorrelation patterns. The results reveal a slight positive spatial autocorrelation, indicating that tree-planting efforts are somewhat clustered in areas with varying levels of fire severity. Significant overlaps are found between tree-planting initiatives and high or very high fire-severity zones, suggesting a potential alignment with fire-risk mitigation strategies. Despite these insights, the analysis is constrained by such limitations as high neighbor counts in the Inverse Distance method and data gaps in the Fixed Distance Band method. Future research should address these limitations by refining analytical parameters and incorporating dynamic data to better understand and optimize the spatial relationship between reforestation efforts and wildfire risks. The findings offer valuable implications for improving tree-planting strategies and enhancing environmental sustainability in the context of wildfire management.**

Keywords—Wildfire; Severity; Spatial Autocorrelation.

I. INTRODUCTION

California, known for its vast landscapes and diverse ecosystems, faces an escalating threat from wildfires [1]. These destructive events have increased in frequency and intensity, posing significant risks to communities, wildlife, and natural resources. Understanding the causes and consequences of wildfires is crucial for developing effective mitigation strategies. One emerging area of interest is the potential correlation between tree-planting initiatives and wildfire outbreaks.

Tree planting, often promoted to combat climate change and enhance urban environments, may have unintended consequences if not properly managed. Recent studies, such as those conducted in Chile, suggest that reforestation efforts without adequate consideration of species selection and environmental impact can exacerbate fire severity [2]. Despite these insights, California lacks comprehensive protocols and checks for tree-planting projects, raising concerns about their role in the state's growing wildfire crisis [3].

This geographic information system project aims to explore the spatial relationship between tree-planting initiatives and wildfire outbreaks in California. By integrating data and employing advanced spatial analysis techniques in ArcGIS Pro, including Global Moran's I, K-Nearest Neighbors, and Fixed Distance Band methods, we sought to uncover patterns and potential correlations. The project leverages data on severe-fire areas and urban tree-planting initiatives to identify hotspots and assess the influence of reforestation on fire severity.

Our findings provide valuable insights into the environmental effects of tree-planting initiatives and inform policy recommendations for sustainable reforestation practices in California. Ultimately, this research strives to contribute to a better understanding of wildfire dynamics and support efforts to mitigate their impact on communities and ecosystems.

Section 2 reviews the relevant literature. Section 3 lays out the methods we used in our study. Section 4 discusses the results of our study. Section 5 concludes.

II. LITERATURE REVIEW

California's history with wildfires is both extensive and devastating, marked by significant events that have shaped the state's landscape and policies. In recent years, several wildfires have set new records for size, destruction, and fatality rates, highlighting the increasing severity of these events. The August Complex Fire of 2020, the Dixie Fire of 2021, and the Camp Fire of 2018 are among the most catastrophic in terms of acres burned, structures destroyed, and lives lost (Figure 1). These fires illustrate a troubling trend: wildfires in California are becoming more frequent and intense, largely due to human activities and climate change. The article in [1] provides a comprehensive overview of California's wildfire history, detailing the 20 largest, deadliest, and most destructive fires. It underscores the critical need for informed strategies to prevent and mitigate future wildfires. The data presented, including annual fire counts, acres burned, and suppression costs, offer valuable insights into the escalating wildfire crisis and the urgency of addressing it. The escalating wildfire crisis in California is not an isolated phenomenon; similar trends have been observed globally, notably in Chile.

In examining global reforestation patterns, researchers have identified significant trends and outcomes with broad implications for environmental policy and biodiversity. Investigations into these patterns have revealed that a substantial portion of new forests will likely consist of plantations rather than natural, diverse forests. This shift towards monoculture plantations is largely driven by economic incentives and the demand for specific types of wood and agricultural products. In Chile, for instance, reforestation efforts have led to an increase in tree cover, yet this has come at the cost of biodiversity [2]. The dominance of monoculture plantations has reduced the variety of species within these forests, undermining the ecological balance and resilience of the region's natural habitats.

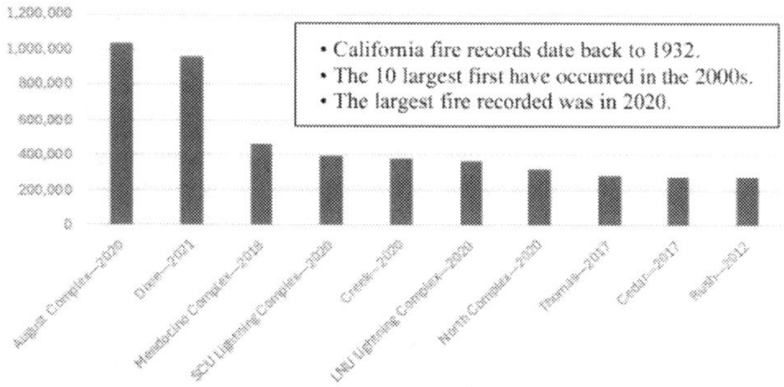

Figure 1. Ten Largest Recorded Fires in California.

While reforestation offers amazing benefits when implemented properly—such as carbon sequestration, soil stabilization, and habitat restoration [4]—Chile's approach has highlighted some critical pitfalls. The country experienced its worst fire season in history in 2024, and experts believe this is largely due to the extensive planting of monoculture forests, which are more susceptible to fire. These findings underscore the need for reforestation strategies that prioritize ecological diversity and sustainability to mitigate the negative impacts on biodiversity and reduce the risk of severe wildfires [2]. The lessons learned from Chile's reforestation challenges underscore the importance of implementing strategic and ecologically sound tree-planting efforts. Poorly designed campaigns could cause more harm than good [3].

Jordan [3] raises critical concerns about large-scale tree-planting campaigns and afforestation subsidies. It highlights that 80% of commitments to the Bonn Challenge, a global initiative aimed at restoring 350 million hectares of degraded and deforested lands by 2030 [5], involve planting monoculture tree plantations, using limited tree mixes focused on specific products like fruit or rubber, and prioritizing plantations over natural forest restoration. These practices might not only fall short of climate goals, but also exacerbate biodiversity loss. The analysis of Chile's Decree Law 701, a significant afforestation program, underscores how poorly designed subsidies can lead to negative outcomes by replacing biodiverse, carbon-rich native forests with less effective plantations [2]. To ensure that tree-planting efforts contribute meaningfully to climate and conservation goals, policies must enforce strict guidelines to prevent the conversion of natural forests into monoculture plantations, thereby promoting genuine ecological restoration and preserving biodiversity. The United States has designated 17 million hectares for restoration while avoiding key biodiversity areas. This trend underscores the need for critically evaluating and improving implementation methods, as seen in California, where current tree-planting strategies do not seem to favor the preservation of natural biodiversity.

Urban tree-planting programs in Los Angeles have been criticized for lacking clear environmental criteria to guide tree selection [6]. This oversight can result in choices that do not support local biodiversity or environmental health. Furthermore, there is no systematic monitoring of the effects of tree planting on the city's environment, making it difficult to assess the success or failure of these initiatives. The Mayor's Tree L.A. (MTLA) program exemplifies these shortcomings, as it lacks a comprehensive, identifiable plan for achieving its goals. Instead, it tends to operate opportunistically, planting trees where partnerships can be established rather than following a strategic, science-based approach [6]. This method may lead to suboptimal outcomes for the urban ecosystem and reduce the overall effectiveness of tree-planting efforts in contributing to urban resilience and sustainability.

III. Methods

Our methods included data selection, spatial autocorrelation analysis and hotspot analysis.

A. Data Selection and Acquisition

We used two key data layers for this analysis. The first layer, detailing fire-hazard severity, was obtained from CAL FIRE's official database [7] and categorizes fire-hazard zones into three classes: moderate, high, and very high. This classification aids in understanding the varying degrees of fire risk across different California regions. The second layer comprises urban tree-canopy data from the U.S. Department of Agriculture [8], created by EarthDefine based on 2018 National Agriculture Imagery Program (NAIP) aerial imagery and light distancing and ranging data collected by the U.S. Geological Survey. The NAIP imagery, captured during the growing season, includes four spectral bands (red, green, blue, and near-infrared) at a 60-cm spatial resolution. Together, these layers offer a comprehensive view of both fire-hazard risks and the extent of urban tree cover, which is crucial for evaluating the intersection of fire management and urban forestry efforts.

B. Analysis Phase

The spatial correlation analysis in ArcGIS Pro begins with the preparation of data layers, which includes loading tree-planting areas and fire-severity zones. These layers are fundamental to understanding the geographic relationship between tree-planting efforts and fire-severity levels. Initially, overlay analysis is conducted using the Intersect tool, combining the tree planting and fire-severity layers to create a new layer that highlights intersecting polygons. These polygons represent areas where tree-planting initiatives coincide with different fire-severity zones. Calculating the area of these intersected polygons provides a quantitative measure of the overlap between tree-planting efforts and fire-prone areas.

Next, the analysis summarizes these areas by fire-severity category. The Summary Statistics tool calculates the total area of tree-planting zones within each fire-severity category, producing a summary table (Table 1). This table is crucial as it breaks down the data into understandable segments, showing the extent of tree-planting efforts in areas with moderate, high, and very high fire risk. Calculating the percentages of these areas relative to the total tree-planting area helps illustrate the proportion of tree-planting efforts that fall within each fire-severity category, providing a clearer picture of whether tree-planting initiatives are strategically aligned with regions of higher fire risk.

TABLE I. ARCGIS SUMMARY TABLE

Object ID	Hazard Class	Incidents	Intersect area (m^2)
2	Moderate	784	690,518,868.872831
1	High	584	549,820,249.080611
3	Very high	1,130	895,145,371.297078

The summary table of the intersected areas provides insights about the distribution of tree-planting efforts within different fire-severity zones, along with the frequency of intersected polygons. For instance, there are 584 polygons intersecting with high fire-severity zones, covering a total area of 549,820,249.08 square meters.

These findings are significant as they illustrate both the frequency and proportion of tree-planting efforts located within different fire-severity zones. By calculating the percentages of these areas relative to the total tree-planting area, the analysis provides a clearer picture of whether tree-planting initiatives are strategically aligned with regions of higher fire risk. The data indicates that a notable portion of tree-planting efforts is concentrated in areas with very high fire risk, suggesting a potential alignment with fire-mitigation strategies. However, the presence of tree planting in moderate and high fire-severity zones also underscores the need for careful planning and management to optimize the benefits of these initiatives in reducing fire risks and promoting environmental sustainability.

C. Spatial Autocorrelation: Global Moran's I Tool

Inverse Distance

We conducted the initial run of the Global Moran's I tool in ArcGIS Pro using the "Intersect_Fire_Tree" layer, focusing on the "Intersect_Area" field (Figure 2). This analysis aimed to assess the spatial autocorrelation of tree-planting and wildfire zones in California. The parameters included the conceptualization of spatial relationships using the Inverse Distance method, with distance measured in Euclidean distance and standardized by Row.

The results indicated a Moran's Index of 0.018046, suggesting a slight positive spatial autocorrelation. The expected index was −0.000400, indicating that any observed clustering is due to random chance. The variance was calculated at 0.000050. The *Z*-score of 2.607674 and the *p*-value of 0.009116 both pointed to a statistically significant clustering pattern, with the *p*-value indicating a less than 1% probability that the observed pattern is due to chance.

However, several warnings were noted during the analysis. The default neighborhood search threshold was set at 151,988.0411 square meters, and at least one feature had over 1,000 neighbors. This high number of neighbors raised concerns about the accuracy of the spatial relationship, indicating that the threshold might need adjustment to refine the analysis and ensure more precise results.

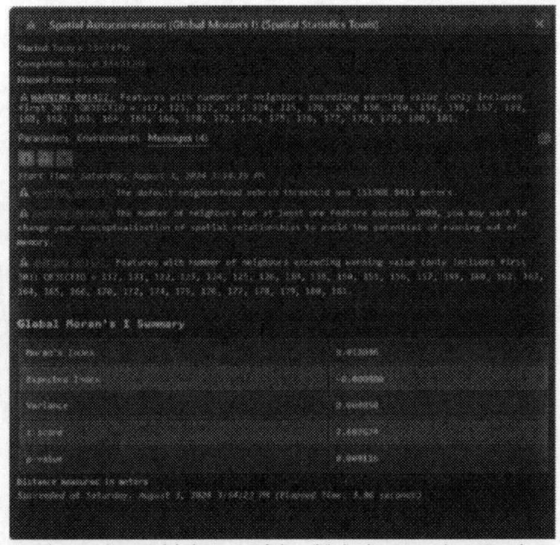

Figure 2. Initial Run of the Global Moran's I Tool.

K-Nearest Neighbor

In the second execution of the Global Moran's I tool, we conducted the analysis using the K-Nearest Neighbors method with a parameter setting of 15 neighbors (Figure 3). The results revealed a Moran's Index of 0.019387, indicating a slight positive spatial autocorrelation. This suggests that areas with higher values of "Intersect_Area" are somewhat clustered together rather than being randomly distributed. The Expected Index was −0.000400, the value one would expect if there were no spatial autocorrelation. The *Z*-Score of 3.120514 and the *p*-value of 0.001805 both signify that this observed pattern is statistically significant. The high *Z*-Score indicates that the Moran's Index is significantly different from what would be expected under the null hypothesis, and the low *p*-value supports the conclusion that the observed clustering is unlikely to be due to random chance. The choice of K-Nearest Neighbors with 15 neighbors allowed for a detailed capture of local spatial patterns, highlighting a notable tendency for similar values to be spatially grouped together.

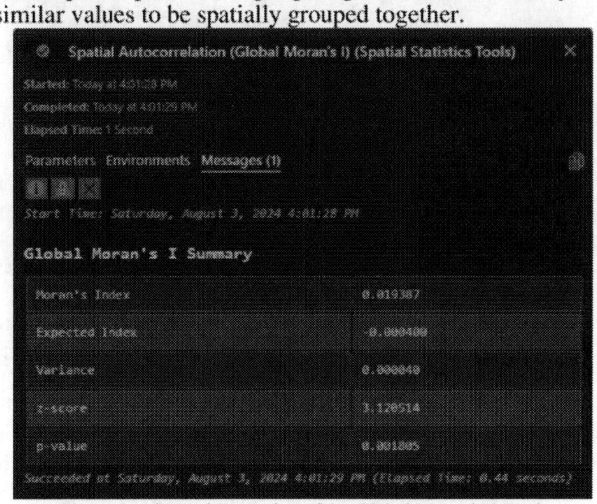

Figure 3. Second Execution of the Global Moran's I Tool.

Fixed Distance Band

In the third execution of the Global Moran's I tool, we used the Fixed Distance Band method with a threshold distance of 50,000 meters (Figure 4). This approach defines spatial relationships based on a fixed radius of 50 kilometers, meaning that only features within this distance from each other are considered neighbors. The results showed a Moran's Index of 0.019167, indicating a slight positive spatial autocorrelation. This suggests that areas with higher values of "Intersect_Area" are somewhat clustered together. The Expected Index was −0.000400, the value anticipated under the null hypothesis of no spatial autocorrelation. The *Z*-Score was 5.641083, a high value that points to a statistically significant deviation from the expected index, while the *p*-value was extremely low, virtually zero, underscoring the significance of the observed clustering. However, warnings indicated that five features had no neighbors within the defined distance, which could potentially affect the validity of the analysis for those features. However, the results indicate that the spatial pattern of "Intersect_Area" values is significantly clustered, demonstrating that similar values tend to group together within 50 kilometers.

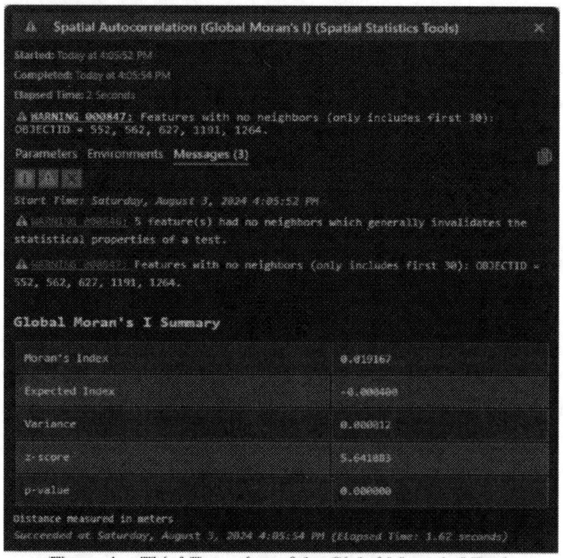

Figure 4. Third Execution of the Global Moran's I Tool.

D. Hotspot Analysis

We conducted a hotspot analysis using ArcGIS Pro to identify fire-severity levels within urban tree-planting areas. The analysis involved calculating *Z*-scores to determine areas of high and low fire intensity. We then created a map to visually represent these severity levels, displaying a range of severity from moderate to very high.

The hotspot analysis illustrated in the map (Figure 5A) demonstrates that the most intense fire-severity areas are concentrated in Southern California. This pattern aligns with other findings [6], which highlight the region's elevated fire risk. The analysis indicates that these areas, marked by high *Z*-scores, correspond to regions with significant overlap between urban tree-planting initiatives and high fire-severity zones. This intense clustering may be linked to the lack of specific environmental criteria guiding tree selection in such programs as the MTLA initiative. The absence of a science-based approach and systematic monitoring in these programs potentially exacerbates fire risk, contributing to the high severity levels observed. This observation underscores the need for more strategic and evidence-based tree-planting practices to mitigate fire risks and enhance urban resilience.

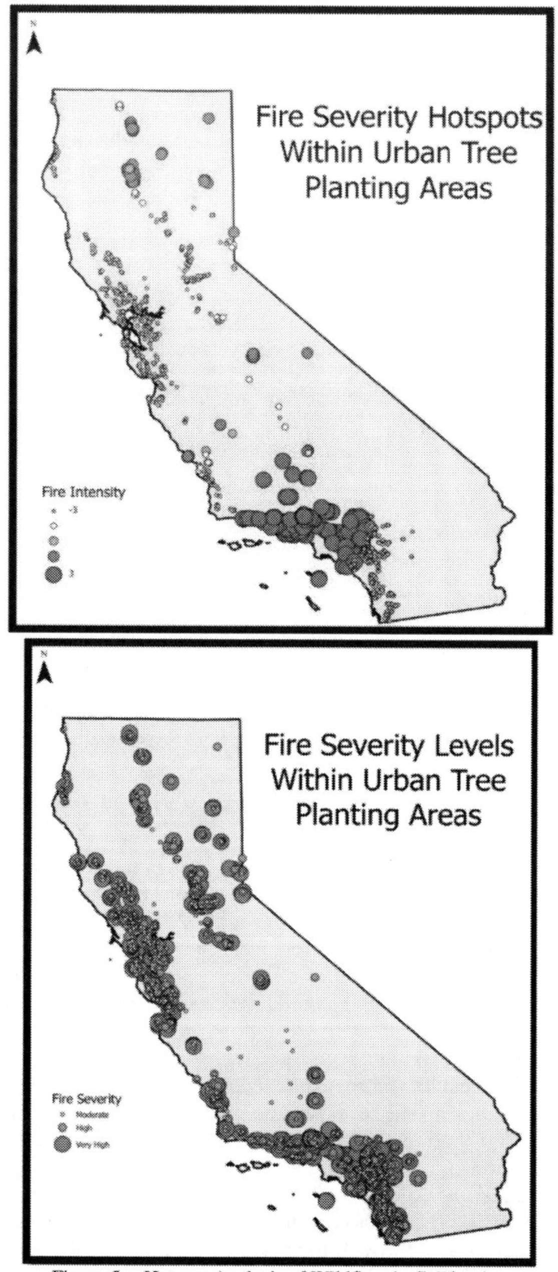

Figure 5. Hotspot Analysis of Wildfires in California.

The second map (Figure 5B) displays circles of three different sizes representing moderate, high, and very high fire-severity levels in areas where the fire-severity layer intersects with the tree-planting layer. While the map shows that high severity is widespread across all intersected areas, the hotspot analysis revealed larger circles concentrated specifically in the Los Angeles area. This discrepancy can be attributed to the spatial clustering identified in the hotspot analysis, which highlighted that the most intense fire severity is significantly more pronounced in certain regions, such as Los Angeles, compared to other areas.

The larger circles in the hotspot analysis represent extreme severity levels, indicating that while high-severity zones are common, Los Angeles exhibits particularly severe fire conditions. This finding is significant as it emphasizes the need for targeted fire-management strategies in high-risk areas. The broader map provides context by showing that high-severity zones are widespread, but the hotspot analysis's focus on Los Angeles underscores the critical nature of fire-resilience efforts in this region, where the intensity of fire activity is notably higher.

IV. Results and Discussion

A. *Results*

We used several methods of spatial correlation analysis in ArcGIS Pro to assess the relationship between tree-planting areas and fire-severity zones. Initially, the data layers for tree planting and fire severity were overlaid using the Intersect tool, creating a new layer that highlighted intersecting polygons. The analysis quantified the overlap between tree-planting initiatives and different fire-severity zones. The summary statistics revealed that the most significant overlap was in very high fire-severity zones (Table 1).

All three runs of the Global Moran's I tool, using the Inverse Distance, K-Nearest Neighbors, and Fixed Distance Band methods yielded Moran's Indexes notably different from the expected value, suggesting a slight positive spatial autocorrelation. The high Z-Scores and low p-values for these indexes indicate that the observed clustering was statistically significant. However, some warnings raised concerns about various issues, suggesting that the parameters may need adjustment.

B. *Discussion*

The results from the spatial-correlation analysis underscore a consistent pattern of positive spatial autocorrelation across different methods. The Intersect tool revealed that a significant portion of tree-planting efforts overlaps with high and very high fire-severity zones, suggesting a potential alignment with fire-risk mitigation strategies. The Global Moran's I analyses using different methods—Inverse Distance, K-Nearest Neighbors, and Fixed Distance Band—each indicated slight positive clustering of tree-planting areas. The high Z-Scores and low p-values across these methods confirm that the observed clustering is statistically significant and not due to random chance.

The Inverse Distance method raised concerns about potential memory issues due to a high number of neighbors, indicating that adjustments might be necessary for future analyses. The K-Nearest Neighbors method provided a detailed view of local spatial patterns, while the Fixed Distance Band method showed a robust clustering effect, despite some limitations related to features with no neighbors. Overall, these findings suggest that tree-planting initiatives are strategically aligned with areas of higher fire risk, though the analysis also highlights the need for careful consideration of spatial relationships and potential data limitations. The results provide valuable insights for planning and managing tree-planting efforts in the context of wildfire risk, emphasizing the importance of addressing spatial patterns to optimize environmental sustainability and fire-risk reduction.

V. Conclusion and Future Work

The spatial-correlation analysis conducted using ArcGIS Pro provides valuable insights into the relationship between tree-planting areas and fire-severity zones in California. The findings consistently indicate a slight positive spatial autocorrelation, suggesting that tree-planting efforts are somewhat clustered in areas with greater fire severity. The analysis using the Intersect tool revealed a significant overlap of tree-planting initiatives with high and very high fire-severity zones, pointing to a potential strategic alignment with fire-risk mitigation efforts. The results from the Global Moran's I tool, employing the Inverse Distance, K-Nearest Neighbors, and Fixed Distance Band methods, all corroborate the presence of significant clustering patterns. These results underscore the importance of considering spatial relationships in the management and planning of tree-planting initiatives to effectively address fire risks and promote environmental sustainability.

A. *Limitations*

Despite the significant findings, this analysis is subject to several limitations. The Inverse Distance method raised concerns due to a high number of neighbors, which could affect the accuracy and efficiency of the analysis. Additionally, the Fixed Distance Band method encountered issues with some features having no neighbors within the specified distance, potentially impacting the validity of the results for those features. The presence of such anomalies suggests that the analysis might be influenced by data gaps or parameter settings that may not fully capture the spatial relationships. Furthermore, the static nature of the distance thresholds and neighbor settings may not account for the dynamic and complex nature of fire-severity and tree-planting patterns.

B. *Future Work*

Future research should address the limitations identified in this analysis by refining the parameters and methods used. Adjusting the neighborhood search thresholds and exploring alternative conceptualizations of spatial relationships could enhance the accuracy and robustness of the spatial autocorrelation results. Additionally, incorporating more dynamic and context-specific data, such as real-time fire-risk assessments and variable tree-planting practices, could provide a more comprehensive understanding of spatial patterns. Future work might also consider expanding the analysis to include other environmental factors or geographic regions to validate the findings and develop more targeted strategies for fire-risk mitigation and tree-planting initiatives. Integrating these improvements will help optimize the effectiveness of tree-planting efforts and better inform environmental-management practices.

REFERENCES

[1] Western Fire Chiefs Association. "History of California Wildfires." WFCA, 17 Nov. 2022. [Online]. Available from: https://wfca.com /wildfire-articles/history-of-california-wildfires/ (retrieved Feb, 2025).

[2] "When Planting Trees Threatens the Forest." News.stanford.edu. [Online]. Available from: https://news.stanford.edu/stories/2020/06 /planting-trees-threatens-forest (retrieved Feb, 2024).

[3] "Poorly Designed Tree-Planting Campaigns Could Do More Harm than Good." Stanford Woods Institute for the Environment, Stanford University, 2020. [Online]. Available from: https://woods.stanford.edu /news/poorly-designed-tree-planting-campaigns-could-do-more-harm -good (retrieved Feb, 2024).

[4] S. C. Cunningham et al., "Balancing the environmental benefits of reforestation in agricultural regions. Perspectives in Plant Ecology." Evolution and Systematics, vol. 17(4), pp. 301–317, 2015.

[5] "Bonn Challenge." Bonn Challenge. [Online]. Available from: https:// www.bonnchallenge.org (retrieved Feb, 2024).

[6] S. Pincetl, T. Gillespie, D. E., Pataki, S. Saatchi, and J. D. Saphores, "Urban tree planting programs, function or fashion? Los Angeles and urban tree planting campaigns." GeoJournal, vol. 78, pp. 475–493, 2013.

[7] "Fire Hazard Severity Zone Viewer—California Open Data." Ca.gov, 2024. [Online]. Available from: https://data.ca.gov/dataset/fire-hazard -severity-zone-viewer (retrieved Feb, 2024).

[8] "Urban Tree Canopy in California." USDA, 2018. [Online]. Available from: https://www.fs.usda.gov/detailfull/r5/communityforests/?cid =fseprd647442 (retrieved Feb, 2024).

Power Grid Vulnerability and Secure Infrastructure for Energy Transmission

Vivian Sultan
California State University
Los Angeles, CA USA
email: vsultan3@calstatela.edu

Philip Caban
California State University
Los Angeles, CA USA
email: pcaban@calstatela.edu

Samuel Wong
California State University
Los Angeles, CA USA
email: swong100@calstatela.edu

Aaditya Kandel
California State University
Los Angeles, CA USA
email: akandel@calstatela.edu

Juan Talome
California State University
Los Angeles, CA USA
email: jtalome@calstatela.edu

Abstract—**This report explores the correlation between modern grid infrastructure and the prevalence of cyberattacks, focusing on California's efforts to modernize its electric grid. We aim to identify areas that have undergone significant technology upgrades, such as monitoring devices, metering, and electric-vehicle charging stations. Using data from the U.S. Department of Energy and the open-source ArcGIS Data Repository, we conducted a spatial analysis to understand how these modernizations correlate with reported cyberattacks. The findings revealed that while modernization is generally associated with improved resilience, the lack of comprehensive data on modern technologies and monitoring devices limited our ability to validate the correlation conclusively. Our study concludes that more granular and relevant data are essential for accurate analysis, and we recommend policies to enhance grid security, including increased investment in cybersecurity measures and comprehensive data collection.**

Keywords—Security; Electric Utilities; Cyberattacks.

I. INTRODUCTION AND PROBLEM DEFINITION

California's proactive stance in modernizing its electric grid aims to improve efficiency, reduce outages, and integrate renewable-energy sources. The electrical grid is undergoing significant modernization efforts, particularly in California, which leads the nation in integrating such new technologies as smart meters, advanced monitoring devices, and electric-vehicle (EV) charging stations. While these advances promise improved efficiency, reliability, and sustainability, they also introduce new vulnerabilities that could be exploited for cyberattacks. This creates a critical need to understand the correlation between these modern grid components and their susceptibility to cyber threats, and to develop effective policies and strategies to enhance the grid's security and resilience. The study builds on theories of infrastructure resilience and cybersecurity and leverages fundamental approaches from geospatial analysis and cybersecurity risk assessment to explore how infrastructure changes influence attack patterns. This report evaluates the impact of modernization on cyberattack risks and provides recommendations to mitigate these risks.

In this report, Section 2 reviews the literature. Section 3 describes the data selection and acquisition. Section 4 details the system used for the analysis. Section 5 lays out the methods used. Section 6 discusses the results. Section 7 offers some recommendations based on the analysis. Section 8 concludes.

California is at the forefront of integrating new technologies, advanced monitoring devices, sophisticated metering systems, and enhanced EV charging stations. Despite these advancements, a notable lack remains in the area of comprehensive, up-to-date reviews exploring the correlation between these modernizations and vulnerability to cyberattacks. This literature review is essential to understanding and mitigating the risks associated with advanced technologies. Previous studies have analyzed the general impacts of modernization on grid resilience, but they have often lacked a focus on the specific correlation between modern grid infrastructure and cyberattacks.

II. LITERATURE REVIEW

Several studies have explored the relationship between grid modernization and cybersecurity. For instance, research has emphasized the critical role of smart grids in enabling two-way communication and real-time monitoring, which significantly enhances operational efficiency [1]. However, these advancements also make grids more susceptible to cyber threats, as highlighted by studies on the vulnerabilities of smart grid components like control centers and substations [2]. Additionally, the U.S. Department of Energy has underscored the importance of cybersecurity measures in protecting modernized grids, citing the increasing frequency and sophistication of cyberattacks [3].

A. *Limitations of Existing Solutions*

While these studies provide valuable insights, they often lack a comprehensive focus on the correlation between specific modernization efforts and cyberattack patterns. For example, most research has concentrated on general vulnerabilities without delving into how particular technologies, such as EV charging stations or advanced metering systems, contribute to these risks [4]. Furthermore, the absence of granular, high-quality data has been a recurring limitation, making it challenging to validate findings conclusively [5].

B. *Future Research*

To address these gaps, future studies should aim to

1. Collect and analyze more granular data on modern grid technologies and their vulnerabilities.
2. Develop robust cybersecurity frameworks tailored to the unique challenges posed by grid modernization.
3. Investigate the role of policy interventions in mitigating cyber risks.

C. *Purpose of This Study*

Building on the existing body of work, this study aims to fill the identified gaps by conducting a spatial analysis of California's modernized grid infrastructure. By leveraging data from the U.S. Department of Energy and the ArcGIS Data Repository, the research seeks to establish a more nuanced understanding of the correlation between modernization efforts and cyberattacks. The goal is to inform policy recommendations that enhance grid security and resilience.

Here's a breakdown of the approach before going into specifics of data selection and acquisition.

1. Objective-Oriented Focus: The central goal was to examine the correlation between electric grid modernization and cyberattacks. This required obtaining datasets that offered both breadth (covering diverse modernization aspects) and depth (granular data for detailed analysis).
2. Practical Considerations: To optimize research within time and resource constraints, emphasis was placed on leveraging readily available and processed datasets. This approach aimed to reduce inefficiencies like excessive data cleaning and processing, which can consume substantial time without guaranteeing usable results.
3. Addressing Data Scarcity: Recognizing gaps in existing data—specifically on emerging technologies like EV charging stations and advanced monitoring devices—alternative strategies were explored. These included using proxy data, expert consultations, and aggregating insights from multiple sources. However, these strategies were limited by their inability to fully bridge the data gaps.
4. System Compatibility: The choice of tools and methodologies was influenced by the compatibility of available data with spatial-analysis and visualization software like ArcGIS and Excel. Computational limitations (e.g., lack of high-end hardware) also shaped the approach, leading to creative solutions like combining maps and tables for more comprehensive analysis.
5. Flexibility and Iteration: Acknowledging the dynamic nature of research, the methodology incorporated iterative steps for refining data acquisition and visualization. This iterative framework allowed the researchers to adapt to the challenges posed by incomplete data and computational constraints.

This high-level framework underscores the balanced approach to addressing the practical and analytical challenges, laying the groundwork for the specifics that follow in data selection and acquisition.

III. Data Selection and Acquisition

Our primary focus when selecting and acquiring datasets on cyberattacks and electric charging stations was ease of use. We prioritized readily available processed data to minimize wasted time. The biggest challenge was the limited availability of specific data on new technologies, monitoring devices, metering systems, and system vulnerabilities. This scarcity is likely due to underreporting or limited awareness.

Fortunately, we were able to leverage clean, processed, and open-source datasets from the official ArcGIS website. These datasets were readily available in ArcGIS Project format. Time constraints and the limited availability of relevant data hindered our ability to draw definitive conclusions about the impact of cyberattacks on charging stations. This, in turn, limited the analysis tools we could employ. Cleaning raw data is time consuming and may yield unusable results. To address these limitations, we considered such alternative approaches as supplementing with proxy data or seeking expert consultations. However, these methods would not fully resolve the data gaps.

One alternative data acquisition method involved importing a CSV file (worksheet) from the US Department of Energy into ArcGIS. This file contained the locations of alternative fuel stations in California, including EV stations (Figure 1). The cyberattack dataset, collected from the open-source ArcGIS hub, included historical cyberincident data along with longitude, latitude, region (Figure 2), city, time zone (Figure 3), and ISP information. This dataset proved to be the most impactful and valuable for our research purposes.

IV. System

Conducting spatial analysis, mapping, and other forms of data visualization demands significant computing power and specialized software. Much of our analysis relied on mobile workstations or laptops equipped with ArcGIS and Excel. The absence of a dedicated graphics card and the inherent limitations of laptops limited our data analysis. Additionally, the data acquired was insufficient for further analysis.

The same data constraints impacted our software choices. ArcGIS proved inadequate for comprehensive spatial analysis, yielding inconclusive results. Consequently, we used Excel, importing attribute tables from ArcGIS into pivot tables. This approach of using both software products provided more insights than the original maps. Subsequent sections of this paper showcase charts and graphs generated from pivot-table analysis, revealing previously undetected patterns.

V. Methodology

We conducted spatial analysis to identify patterns and relationships between modernization features and attack frequencies; the data for this analysis was provided by Arc GIS.com and the U.S. Department of Energy. The analysis shows a correlation between the modernization of the electrical grid and cyberattack incidents and hotspots with higher levels of cyberattacks. The Data Overlay tool was used to visualize the overlap between grid modernization efforts and cyberattack incidents, and statistical methods were used to identify patterns and relationships between modernization features and attack frequencies. We employed such geographic information system (GIS) functions and operations as the *create new map layer* feature, which helped by combining cyberattack events and EV charging station locations to create a new map layer named MODERN POWER GRID/CYBERATTACK LOCATION. Another ArcGIS Pro tool used was the Merge tool to combine two or more features on the same layer into a new feature. Bar charts

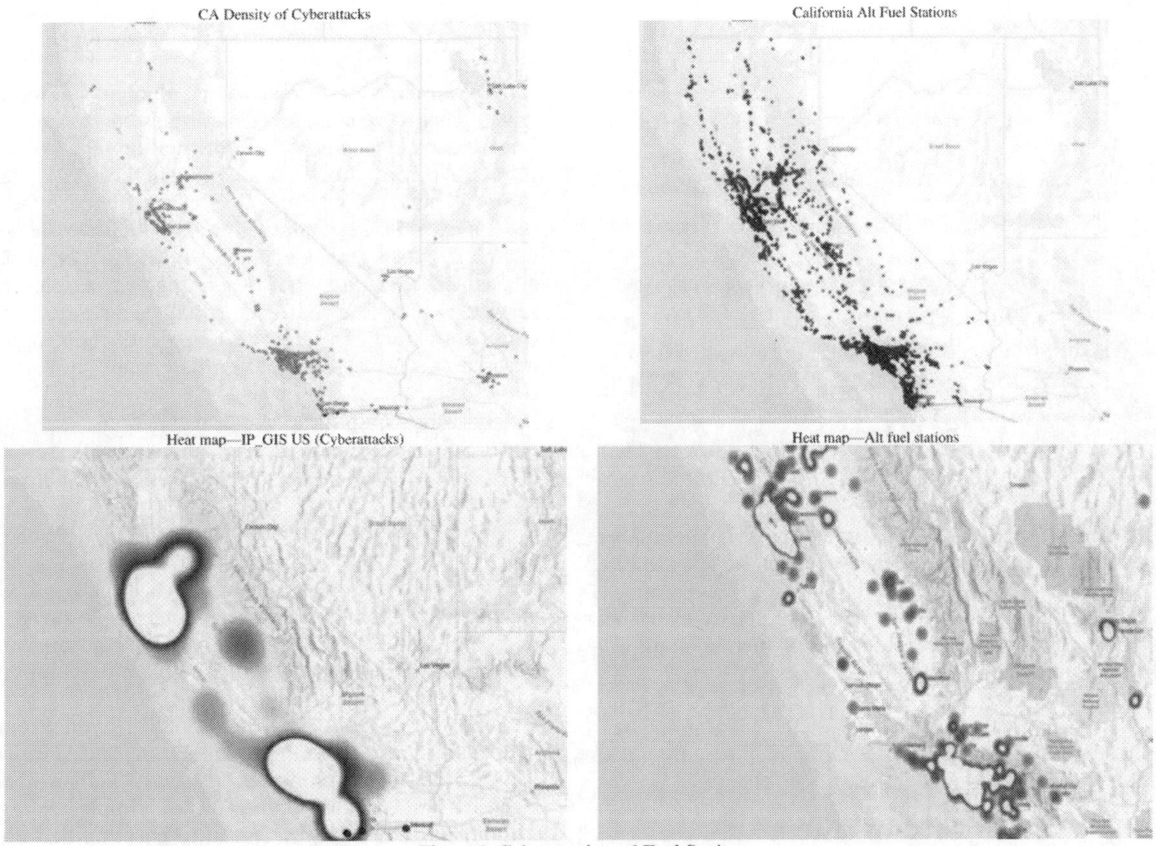

Figure 1. Cyberattacks and Fuel Stations.

were also used to show types of EV ports at different charging stations and cyberattacks by city, revealing which cities have higher levels of cyberattacks (Figure 4).

VI. RESULTS/DISCUSSION

After performing our ArcGIS Pro analysis, we found specific hotspots where charging stations are more susceptible to being hacked, putting people's personal data in danger (Figure 5). The outcome of our initial analysis indicated a general pattern where modernized areas with advanced technologies exhibited varying degrees of cyberattack susceptibility. However, the results were inconclusive due to insufficient data on specific modernization aspects, such as monitoring devices and metering systems. These modernization aspects are important because their absence can expose sensitive such information as credit-card information or vehicle data, compromising people's privacy. The findings from this research project can raise awareness of this issue and prompt electrical companies with vulnerable locations in the EV charging infrastructure to enhance their equipment and protect against cyberattacks. The main problem we encountered was the lack of relevant data because little research and data is available on this topic. Time constraints were also a limitation because of the small timeframe given in class.

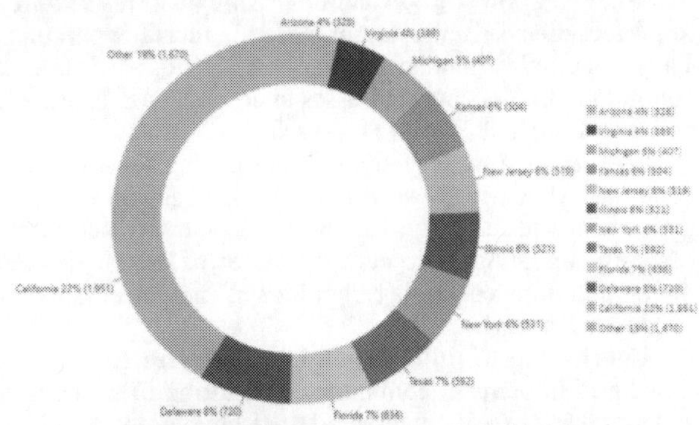

Figure 2. Percentage of Cyberattacks by Region.

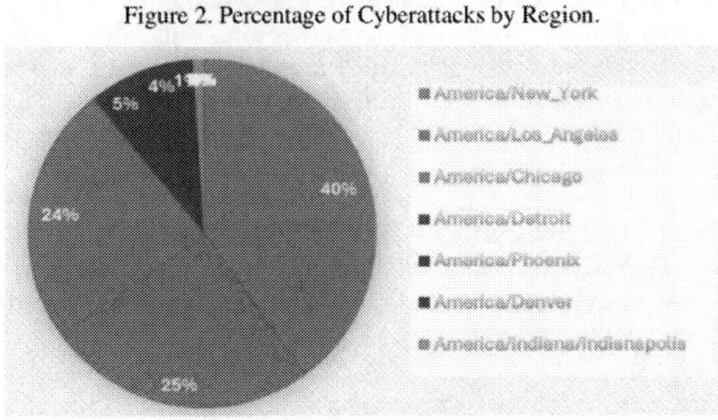

Figure 3. Cyberattacks by Time Zone.

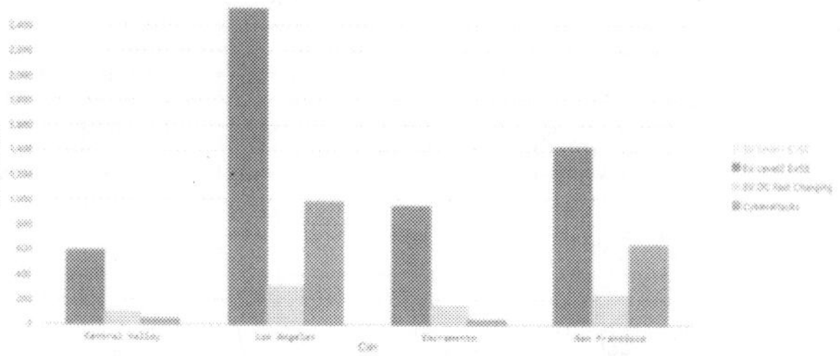

Figure 4. EV Charging and Cyberattacks by Region.

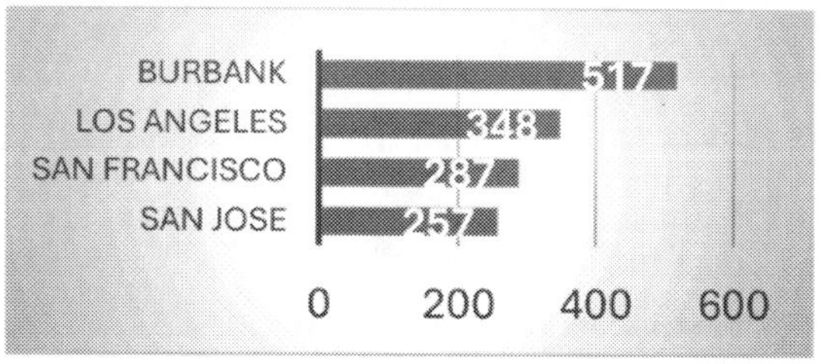

Figure 5. Most Attacked California Cities.

- **Outcomes:** The analysis indicated a general pattern where modernized areas with advanced technologies exhibited varying degrees of susceptibility to cyberattacks. However, the results were inconclusive due to insufficient data on specific modernization aspects, such as monitoring devices and metering systems.
- **Visual Output:** The report includes maps showing modernized grid areas and their correlation with cyberattack incidents, alongside graphs illustrating attack frequencies in relation to infrastructure changes.

VII. Recommendations

What can power grids and other similar infrastructure do to protect their systems and prepare for future cyberattacks? They can employ enhanced cyberattack standards, implement continuous monitoring and assessment, develop stakeholder collaboration, and invest in research.

Enhanced Cybersecurity Standards: Developing or adopting cybersecurity standards or practices tailor-made for the current and future grid technologies and infrastructure by regularly updating IT security policies based on the latest threat intelligence and technological advancements will improve IT security.

Continuous Monitoring and Assessment: How would a smart grid implement continuous monitoring of grid systems and regular assessment of modern technologies to identify vulnerabilities? Teams of network professionals can enforce updated practices through such monitoring systems as a real-time dashboard of smart-grid connected IoT devices. Such a dashboard could provide continuous information for data warehouses to be analyzed for cyberattacks trends and to test the effectiveness of modern security measures before exploits can harm the integrity of a grid's infrastructure.

Enhanced Data Collection: Collect and analyze more data on cyberattack trends and the effectiveness of modern security measures.

Stakeholder Collaboration: Foster collaboration between technology providers, grid operators, and cybersecurity experts to share threat intelligence and best practices.

Investment in Research: Support research and development into new technologies and their security implications to stay ahead of potential threats. Consider how future cybercriminals can gain unauthorized access. For example, outdated system software, insufficient cybersecurity measures, and lack of encryption is enough to risk any highly technical smart grid. Cybersecurity and cybercriminals are constantly changing and innovating, from malware as a service to simple social-engineering techniques. Therefore, continuous research and development, though costly, is necessary to protect the grid in the equally growing field of cybersecurity.

By addressing these recommendations, California can better safeguard its electric grid against evolving cyberthreats and continue leading the nation in modernizing its energy infrastructure.

VIII. Conclusion

California's efforts to modernize its electric grid with new technologies, monitoring devices, metering systems, and EV charging stations represent significant advancements in improving grid efficiency and sustainability. However, these advancements also introduce new cybersecurity challenges. By adopting robust cybersecurity frameworks, enhancing incident response, fostering collaboration, and addressing data gaps, California can reinforce the security and resilience of its electric grid against cyberattacks. Implementing these recommendations will be crucial to ensuring that the state's grid modernization efforts lead to a more reliable and secure energy infrastructure. The inconclusive results regarding the impact of modernization on cyberattack frequencies highlight the need for further research and data collection to validate the effectiveness of security measures and technologies.

References

[1] National Institute of Standards and Technology, "Cybersecurity for smart grid systems," 2023. [Online]. Available from: https://www.nist.gov/programs-projects/cybersecurity-smart-grid-systems. Accessed March 3, 2025

[2] U.S. Department of Energy, "Spotlight: Advancing cybersecurity to strengthen the modern grid," 2021. [Online]. Available from: https://www.energy.gov/sites/default/files/2021/01/f82/OTT-Spotlight-on-Cybersecurity-final-01-21.pdf. Accessed March 3, 2025

[3] Upstream, "Automotive & smart mobility global cybersecurity report," 2025. [Online]. Available from: https://upstream.auto/ty-upstreams-2025-global-automotive-cybersecurity-report. Accessed March 3, 2025

[4] Clarion Energy, "Data is the backbone of grid modernization," Renewable Energy World. 2024. [Online]. Available from: https://www.renewableenergyworld.com/power-grid/smart-grids/data-is-the-backbone-of-grid-modernization. Accessed March 3, 2025

[5] National Association of Regulatory Utility Commissioners, "Cybersecurity baselines for electric distribution systems and DER," 2022. [Online]. Available from: https://www.naruc.org/core-sectors/critical-infrastructure-and-cybersecurity/cybersecurity-for-utility-regulators/cybersecurity-baselines. Accessed March 3, 2025

Analyzing Electric Vehicle Charging Infrastructure Accessibility in Los Angeles, California

Vivian Sultan
California State University
Los Angeles, CA USA
email: vsultan3@calstatela.edu

Oscar Aguillon
California State University
Los Angeles, CA USA
email: oaguillon@calstatela.edu

***Abstract*—This project evaluates the distribution and accessibility of electric vehicle (EV) charging stations in Los Angeles, California, using a variety of geospatial analysis tools. As electric vehicles have become more popular, an accessible charging infrastructure will crucially support and encourage future growth. By performing service-area and closest-facility analyses, this study identifies key areas where EV charging stations are accessible and highlights regions still needing development. The findings emphasize the importance of strategic planning in EV infrastructure to enhance accessibility and promote wider adoption. This research provides insights for urban planners and policymakers to address gaps in the current infrastructure and propose new strategies for future development and demonstrates an analysis model applicable to other areas.**

***Keywords*—LiDAR; deep learning; point cloud; ArcGIS Pro; point classification.**

I. INTRODUCTION

The growing adoption of electric vehicles (EVs) marks a big shift in the automotive industry, led by environmental concerns and technological advances. To accommodate the increasing number of EVs on the road, an accessible charging infrastructure must be developed. The Los Angeles region, known for its high population density and large urban landscape, presents both challenges and opportunities for implementing an EV-charging infrastructure.

California leads the nation in EV adoption rates and demonstrates a strong commitment to reducing greenhouse gas emissions and promoting sustainable transportation solutions. However, achieving these goals requires more than just increasing the number of electric vehicles; a network of easily accessible charging stations for all users, regardless of their location within the city, is crucial.

This project evaluates the current distribution and accessibility of EV charging stations in the Los Angeles area using ArcGIS Pro's geospatial analysis tools. Using the service-area and closest-facility analyses, this study aims to identify areas well served by existing EV charging station infrastructure and those that are underserved and require further development.

Understanding the current spatial distribution of charging stations enables urban planners and policymakers to improve the EV charging network based on real data. An effective charging infrastructure can reduce the range anxiety common among potential EV buyers, encouraging wider adoption and helping the state achieve its environmental goals. Furthermore, insights gained from these analyses can inform future planning efforts, ensuring that investments in EV infrastructure are well targeted and effective in the long run.

This research aims to provide valuable insights into Los Angeles' current EV charging infrastructure, highlight areas for improvement and offering recommendations for future development based on real data. This study's findings will contribute to ongoing sustainable-transportation discussions and EVs' specific role in achieving these targets.

The rest of this report is structured as follows. Section 2 reviews the relevant literature. Section 3 details the methods used. Section 4 outlines the results. Section 5 discusses the findings and their implications. Section 6 concludes.

II. LITERATURE REVIEW

Cui et al. [1] focused on methods for strategically placing EV-charging stations in urban areas, highlighting the importance of considering power-flow constraints and protection-device upgrades to avoid costly future upgrades and ensure efficient power distribution. This analysis helped understand the complexities of charging-station placement and clarified the research goals.

Wen et al. [2] examined global trends and challenges in EV-charging-infrastructure deployment and emphasized the need for strategic planning and energy management to meet increasing EV demand, categorizing different planning models for various regions. This study gave a global perspective on EV infrastructure and its potential benefits for sustainability and grid-strain reduction, helping me understand the research on Los Angeles and such key factors as grid strain.

Furthermore, Azadfar et al. [3] determined optimal EV-charging-station placement using advanced algorithms and models, considering such factors as traffic flow and user convenience. This study gave me insights into advanced analytic techniques for optimal EV charging station placement, helping inform the ArcGIS spatial analysis.

The methodology of Sultan et al. [4], which used geographic information system (GIS) tools to identify gaps in the EV-charging infrastructure, was extremely useful for understanding current infrastructure assessments. Their approach highlighted the need for more stations, but did not incorporate dynamic factors like traffic patterns, inspiring me to explore that avenue myself. We expanded our analysis to include real-time data, aiming to provide a more detailed picture of accessibility and distribution. By incorporating live-traffic impacts on charging-station accessibility, this report builds on this foundation and methodology.

Overall, these studies provided a comprehensive understanding of the EV-charging-infrastructure landscape, highlighting challenges, opportunities, and recommendations. They guided research question formulation and methodology, ensuring the analysis was informed in existing literature while adding meaningful data like live traffic patterns to distinguish the research.

III. METHODOLOGY

This study used a geospatial analysis methodology through ArcGIS Pro to evaluate the current distribution and accessibility of EV charging stations in the Los Angeles region (Figure 1). This approach involved several key components, each designed to assess the current infrastructure and identify areas for potential improvement.

The first phase of the methodology involved initial data collection, which focused on gathering and importing datasets on EV-charging-station locations, road networks—including primary and secondary roads—and demographic characteristics of the region, including live traffic data (Figure 2). The data was sourced from the ArcGIS Living Atlas, ensuring a reliable dataset and an easy framework to import into ArcGIS Pro.

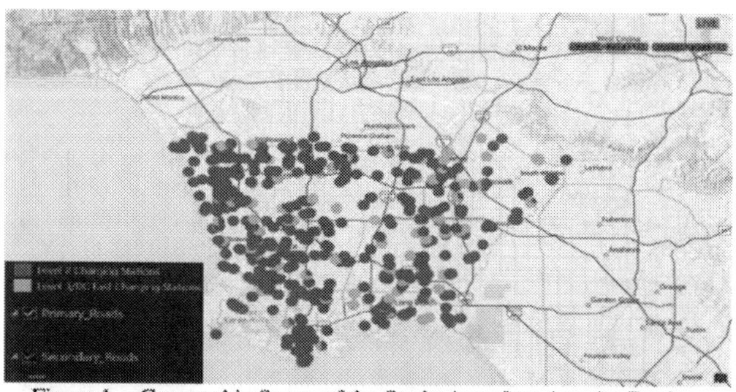

Figure 1. Geographic Scope of the Study Area: Los Angeles Region.

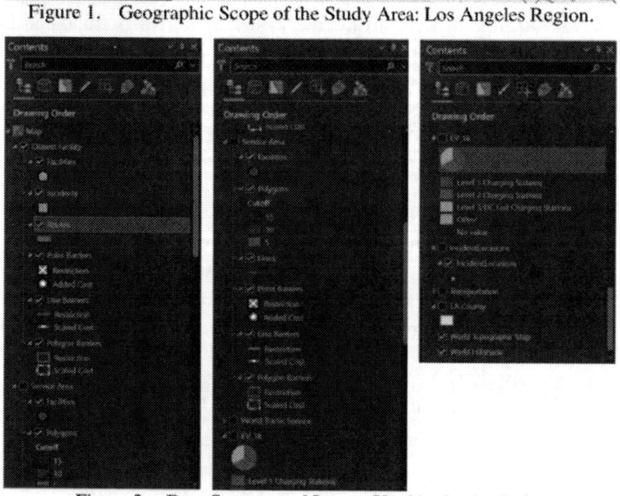

Figure 2. Data Sources and Layers Used in the Analysis.

With the data successfully imported, the second phase centered around using ArcGIS Pro's advanced geospatial tools. Specifically, this project used the Service Area Analysis and Closest Facility Analysis Network Analyst tools. These tools played an integral role in mapping the accessibility and distribution of charging stations. The Service Area Analysis tool was employed to create polygons representing areas within a certain travel time from each charging station. This step was crucial to understanding the reach of each station and highlighting underserved areas that fell outside typical driving distances. It visually identified potential gaps in the infrastructure.

Additionally, we used the Closest Facility Analysis tool to determine the nearest charging station to selected incident points. These points represented such key locations as major transportation hubs, educational institutions, residential areas, and major traffic intersections, providing insights into the practical day-to-day accessibility of charging stations from these locations throughout the region.

An important aspect of the methodology was integrating real-time traffic data into the analysis. Analyzing changing travel conditions enhanced the accessibility assessments, offering a more realistic view of how traffic patterns impact charging stations' accessibility.

The final phase of the methodology involved visualizing and interpreting the results. Thematic maps visually represent the distribution and accessibility of charging stations. These maps facilitated the identification of areas well served by the current EV infrastructure and those requiring further development.

Overall, the various phases of the methodology process for this project facilitated a detailed evaluation of the current EV charging infrastructure in the Los Angeles region. The insights gained from this analysis can help urban planners and policymakers make data-driven decisions to enhance the accessibility of charging stations and ultimately promote wider adoption of electric vehicles.

IV. RESULTS

The analysis of service areas around each EV charging station was organized into three main categories based on travel time: 5 minutes, 10 minutes, and 15 minutes (Figure 3). Each category was visually represented using specific shades of purple, with lighter shades depicting shorter travel times. The 5-minute areas, depicted in light purple, are the most densely covered and indicate regions where an EV charging station is highly accessible. Residents and commuters in these areas can conveniently reach a charging station within a 5-minute drive, ensuring that EV users have easy access to charging facilities for their vehicles.

The 10-minute areas, represented in medium purple, serve as transitional areas with moderate coverage. While they suggest a reasonable level of accessibility, these areas might be at the edge of convenience for those needing quick charges. Residents in these areas may experience slightly longer travel times to access charging stations, which could impact the decision-making of potential EV buyers.

The 15-minute areas, shown in dark purple, indicate zones where accessibility drops significantly. These regions could be considered underserved in terms of rapid access to EV charging facilities, challenging current and potential EV owners. The limited accessibility in these areas suggests a need for expansion of the charging network to better serve the population and encourage broader adoption of electric vehicles.

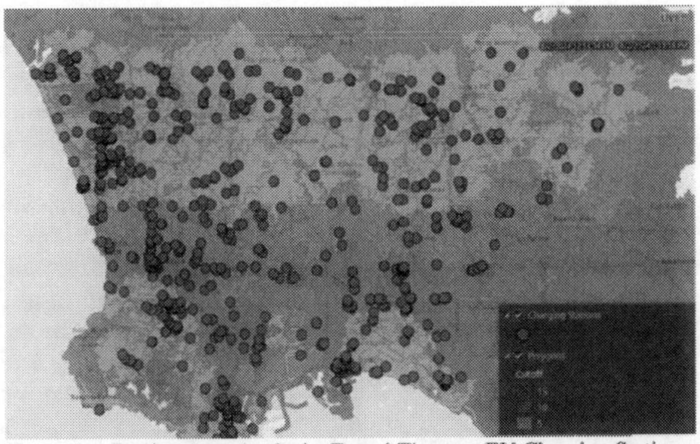

Figure 3. Service-Area Analysis: Travel Times to EV Charging Stations.

The implications of these accessibility levels are significant. The lighter shaded areas, predominantly centered around major urban and commercial hubs, reflect solid infrastructure catering to a high density of EV users. In contrast, the darker shades, particularly found in the outskirts and less urbanized areas, highlight the need for expansion of the charging network. Addressing these underserved areas is crucial to support both potential and existing EV owners and to encourage further growth in EV adoption.

Integrating real-time traffic data provided a dynamic overlay to the service-area polygons (Figure 4). This inclusion offered insights into how regular traffic conditions affect accessibility to charging stations during different times of the day or week. By considering traffic patterns, the analysis goes beyond just geographical proximity, providing a more detailed understanding of accessibility challenges that EV users might face in their daily commutes or travel routines.

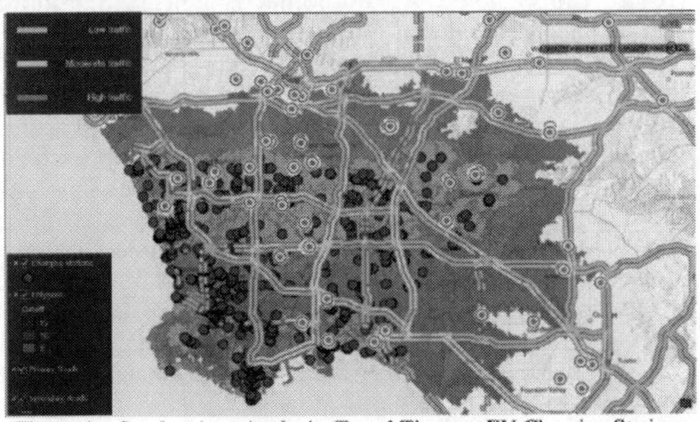

Figure 4. Service-Area Analysis: Travel Times to EV Charging Stations with Live Traffic.

Thematic maps generated from the analysis visually show the service coverage. These maps are instrumental in identifying specific areas where policy interventions, infrastructure enhancements, and targeted investments could significantly improve service reach and, consequently, EV adoption rates. By identifying gaps in coverage, stakeholders can make informed decisions to strategically enhance the EV charging infrastructure, ultimately supporting the transition to more sustainable transportation options.

We conducted the closest-facility analysis to evaluate the accessibility of EV charging stations from specific incident points within both urban and suburban regions (Figure 5). These points are depicted as orange squares on the map, highlighting the proximity to nearby charging stations. This analysis was crucial in determining which areas have sufficient coverage and which require infrastructure improvements. Most designated points, including major transportation hubs like Los Angeles and Long Beach airports, as well as educational institutions such as Cal State Long Beach, are conveniently located within a 5-minute drive from a charging station, demonstrating the great accessibility within the existing EV network.

Incidents

Field: Add Calculate Selection:

	ObjectID *	Shape *	Name
1	9	Point	I-405 & I-105 Interchange
2	16	Point	I-105 & I-710 Major Inte...
3	17	Point	Palos Verdes
4	18	Point	Long Beach Airport
5	20	Point	Hawthorne Residential
6	21	Point	Norwalk Residential
7	23	Point	CSULB
8	24	Point	I-105 & I-110 Major Inte...
9	25	Point	LAX
10	26	Point	Whittier Residential
11	27	Point	Willowbrook Residential
12	28	Point	Dominguez Hills Reside...
13	29	Point	Redondo Beach Residen...
14	[illegible]	[illegible]	[illegible]

Figure 5. Incident Points: Closest-Facility Analysis.

However, certain residential areas like Whittier, Palos Verdes, and Dominguez Hills displayed significant gaps, with no stations within a 5-minute radius (Figures 6 and 7). Particularly in Palos Verdes, disparities in access were evident even within the same city, as one incident point within the city had a charging station within 5 minutes, while another did not. This difference within the same city demonstrates the accessibility challenges EV users might face due to lacking infrastructure. Enhancing the infrastructure in these areas could greatly improve the viability and appeal of EV ownership, potentially fostering wider adoption.

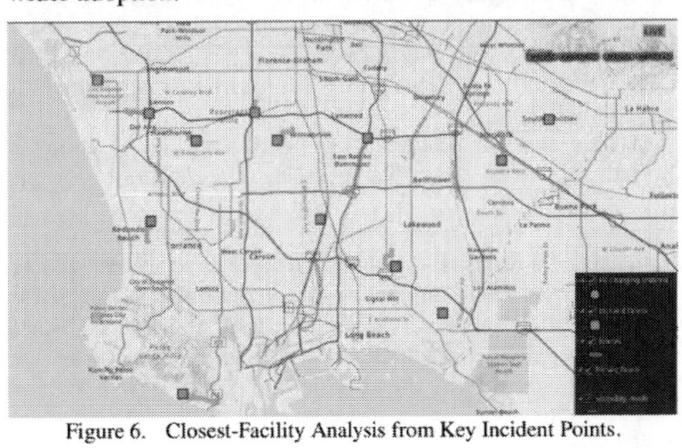

Figure 6. Closest-Facility Analysis from Key Incident Points.

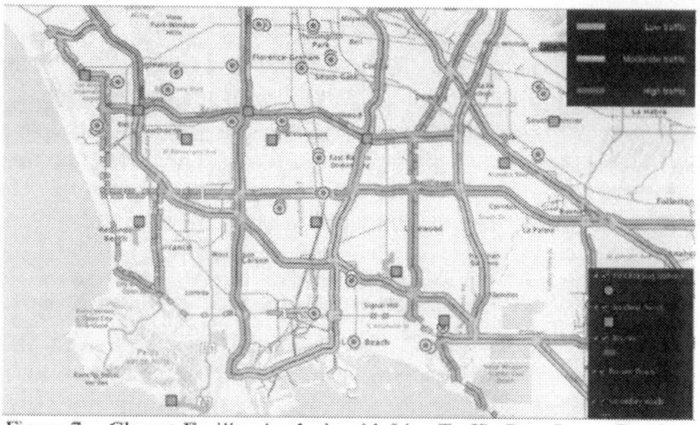

Figure 7. Closest-Facility Analysis with Live Traffic Data Layer Overlay.

The findings from this analysis are crucial for urban planners and stakeholders, indicating a clear need for more equitable charging-station access across all regions to support and encourage the adoption of electric vehicles.

V. DISCUSSION

The geospatial analysis conducted in this study demonstrates the disparities in the accessibility of EV charging stations across the Los Angeles region. The results highlight a need for strategic planning and expansion of the EV infrastructure to support the growing demand for electric vehicles. Specifically, the service area analysis revealed that while major urban centers and commercial hubs have an accessible EV-charging infrastructure, suburban and less urbanized regions face accessibility challenges. This imbalance aligns with previous studies [4], which identified gaps in the EV charging infrastructure and discussed the importance of comprehensive planning to address this. The visual representation through the service-area-analysis tool helped highlight underserved areas, particularly those with 15-minute or greater travel times to the nearest charging station, indicating a pressing need for targeted infrastructure development based on real data.

Using real-time traffic data in the analysis, this study provided a deeper understanding of accessibility challenges than static geographical analysis alone. Incorporating the dynamic nature of traffic conditions allows us to evaluate the practical challenges of reaching charging stations, with traffic being a well-known barrier. Considering such factors in infrastructure planning is crucial as it highlights all the complex elements involved in this type of infrastructure. This approach builds upon the methodology suggested by Azadfar et al. [3], who advocated for incorporating traffic flow and user convenience in determining charging-station placement.

This study's findings have significant implications for urban planners and policymakers. The accessibility gaps identified in areas like Whittier, Palos Verdes, and Dominguez Hills suggest a need for equitable access to EV charging facilities. By addressing these gaps, urban planners and policymakers can potentially reduce the range anxiety among potential EV users and help encourage broader adoption of electric vehicles. The strategic placement of new charging stations in these underserved areas will enhance accessibility, reduce travel times, and align with the planning models discussed by Wen et al. [2]. Additionally, considering the environmental and sustainability goals of the Los Angeles region, a pioneer in EV infrastructure, these recommendations can significantly contribute to expanding the EV charging network, thereby reducing greenhouse gas emissions and promoting cleaner transportation alternatives.

To promote wider EV adoption and support the transition to more sustainable transportation, this study recommends expanding the charging infrastructure by targeting areas identified as underserved, particularly those with limited access within a 15-minute travel time, and prioritizing these locations for new charging-station installations. This study also suggests engaging stakeholders by collaborating with local governments and utility companies to develop a plan for EV-infrastructure expansion that reflects the needs and inputs of these various entities, including stakeholders and investors, to ensure the effective expansion of EV infrastructure.

VI. CONCLUSION

This study has provided an in-depth analysis of the accessibility and distribution of EV charging stations in the Los Angeles region, focusing on key areas that require attention and development. Geospatial analysis tools such as service area analysis and closest facility analysis within ArcGIS Pro identified significant gaps in the current EV infrastructure, particularly in suburban and less urbanized areas. The integration of real-time traffic data into the analysis highlighted the importance of considering dynamic factors when planning for infrastructure development.

This research supports the need for strategic planning and investment to enhance the accessibility of EV charging stations, which will ultimately support the broader adoption of electric vehicles. As California continues to lead the nation in EV adoption, the infrastructure must keep pace with demand to ensure a successful transition. This study's recommendations offer a promising roadmap for urban planners, policymakers, and stakeholders to develop a more equitable and efficient EV charging network. By prioritizing underserved areas and engaging in open discussions with local governments, utility companies, and other stakeholders, Los Angeles can set an example for other regions striving to enhance and develop their EV infrastructure. It is all about planning strategically and using real-world data to inform decisions.

In conclusion, this research provides valuable insights into the current state of Los Angeles' EV-charging infrastructure, as well as practical solutions for its improvement. Real-time data and advanced geospatial analysis can inform decisions to align with the evolving needs of the community and contribute to the broader goal of sustainable urban transportation. Collaboration between public and private entities will be crucial to achieving these objectives and ensuring that Los Angeles continues to lead the EV revolution.

REFERENCES

[1] Y. Cui, J. Xu, Y. Li, and Y. Wang, "Electric vehicle charging station placement method for urban areas." J Urban Plan Dev, vol. 148, no. 2, art. 04022007, 2022. [Online]. Available from: https://doi.org/10.1061/(ASCE)UP.1943-5444.0000720

[2] H. Wen, X. Li, J. Chen, and J. Zhang, "Electric vehicles charging infrastructure demand and deployment: Challenges and solutions." Energy Policy, vol. 158, pp. 112–123, 2023. [Online]. Available from: https://doi.org/10.1016/j.enpol.2021.112123

[3] E. Azadfar, V. Sreeram, and D. Harries, "Electric vehicle charging strategy study and application on charging station placement. Renew Sust Energ Rev, vol. 52, pp. 1497–1507, 2015. [Online]. Available from: https://doi.org/10.1016/j.rser.2015.07.100

[4] V. Sultan et al., "Spatial analysis to identify the need for additional EV charging." In ENERGY 2023: The Thirteenth International Conference on Smart Grids, Green Communications and IT Energy-Aware Technologies (pp. 49–56). International Academy, Research, and Industry Association, 2023.

Potential for Wind Energy on the Fort Bidwell Reservation

Vivian Sultan
California State University
Los Angeles, CA USA
email: vsultan3@calstatela.edu

Hector Tarango
California State University
Los Angeles, CA USA
email: htarango@calstatela.edu

***Abstract*—Tribal lands account for 2.4% of U.S. total acreage and, as of 2021, nearly 8% of its wind energy potential. These lands are commonly rural and historically have had less than average access to electricity. This project's objective was to examine the Fort Bidwell Reservation and locate a site for wind energy generation using ArcGIS' buffer tool. The buffer tool's output boundaries were cross referenced with wind density maps and tribal territory to find a specific location suitable for a wind energy project. The project was technically successful. Results show two appropriate locations, around 100 acres and 200 acres respectively, much larger areas than needed for a wind farm. However, further research is necessary as these two sites are at the outer edges of the reservation and at high elevation. These limitations were outside of the scope of the project and, therefore, it cannot be said with absolute certainty that these two sites are 100% conforming. Furthermore, while an area may be technically suitable, it is also necessary to consult with tribal leadership to ensure the land is not of cultural significance and, therefore, not available for development.**

Keywords*—*Energy Sovereignty; Wind; ArcGIS.

I. INTRODUCTION AND PROBLEM DEFINITION

The power grid faces many new obstacles that range from technological innovation to climate change. The pivot away from fossil fuels and the diversification of the grid through alternative energy sources is central to both challenges. It is understood that rural populations present unpredictable stress on the power grid in times of instability. Shoring up these points of grid weakness can reinforce the grid's overall stability. The native people who inhabit U.S. tribal lands (2.4% of total U.S. acreage [1,2]) are under a particular set of circumstances. Many such communities are underserved, some without access to electricity at all [3], despite them representing nearly 8% of U.S. wind potential [4,5]. By focusing resources on these areas, we can address grid stability, grid diversification, and climate justice all at once. Wind energy generation is a more mature technology today than in the recent past and can be harnessed by these communities to pull themselves out of energy poverty, perhaps even achieving energy sovereignty.

This project's primary objective was to ascertain whether the Fort Bidwell Reservation can host a wind energy farm on its land. Most importantly, the will and desires of the Fort Bidwell Reservation community must be prioritized. As part of the project, other energy project proposals were researched. The goal of locating an appropriate wind farm site is not to be placed over other reasonable solutions available if they present themselves. Tribal land data and wind density maps were cross referenced. ArcGIS Pro's buffer tool created boundaries around watercourses, residential areas and a local airport. We expected that once these restrictions were in place, an appropriate site would be revealed given the area's rich resource and the large size of the reservation.

The Fort Bidwell Reservation tribal leadership is the intended audience of the research as well as government programs designed to diversify the power grid. Energy independence can be a steppingstone to a more educated and healthier community. If a large enough site is found, it may be possible that beyond supporting local infrastructure, energy could potentially be returned to the California grid to provide financial benefits to the people of the Fort Bidwell Reservation. This type of project, if at large enough scale, could also add jobs to the local economy. The benefits of a successful wind farm are many.

Section 2 reviews the literature. Section 3 describes the data selection and acquisition. Section 4 details the system used for the analysis. Section 5 lays out the methods used. Section 6 discusses the results. Section 7 concludes.

II. LITERATURE REVIEW

The topic of the project was selected because it is at the intersection of power diversification, emerging technologies, climate change, and energy justice. All energy producers will need to continue efforts to diversify the power grid to combat the negative impacts caused by climate change and increased demand. This increased demand will be in part due to migration to such newer energy technologies as electric vehicles. Given that native reservations are typically isolated and underdeveloped, it makes them unusually vulnerable to energy instability. However, native lands are generally rich in alternative energy potential and, thus, could be positioned to become energy sovereign soon.

The literature review was focused on two main areas. The first area is wind turbine farming: pros, cons and other relevant information. The review's goal was to determine what requirements must be met prior to project initiation, what exactly qualifies as a good candidate for wind farming and what alternatives to wind farming may be better suited for this specific geographic area. The second area of literature review was the history of renewable energy projects on or near Fort Bidwell and the history of the Fort Bidwell Reservation.

Rediske et al. [6] concisely stated that "wind energy is abundant, costless, widely distributed and its generation is pollution-free." Detailed in this article is a comprehensive guide to evaluating an area to determine if it is suitable for a wind farm project. The first step in such an investigation is accounting for restrictive factors, commonly referred to as exclusion criteria. To maintain a manageable scope for this project, only a limited number of restrictive factors were considered. The article details twenty considerations, but this project was limited to the most relevant five. The factors included road networks, urban areas, airports, wind speed and watercourses. A wind farm needs to be within 500 meters of the main road network [6]. Urban areas should be a minimum of 1,000 meters away from wind farms due to noise and shading generated by the turbine blades. Airports must be at minimum 2,500 meters away from wind farms because electromagnetic radiation from the turbines may interfere with telecommunication networks. One of the most crucial restrictions is wind speed, the site's wind speed must be greater than 7 meters per second and less than 25 meters per second to avoid damage from strong winds. Finally, the last restrictive factor for this project was proximity to watercourses and streams because wind farms should not be within 400 meters of rivers or streams. These restrictions should provide sufficient guardrails when assessing potential sites for a wind farm on the Fort Bidwell Reservation.

The barriers to renewable energy generation must be researched prior to any renewable energy generation project. The realities of historic disenfranchisement and the lingering effects thereof must be addressed to ascertain the feasibility of such a large scale project. U.S. native reservations are under the jurisdiction of the federal government. However, these tribes were historically unable to make use of federal tax credits which greatly incentivize renewable development [7]. Wolfe [8] stated, "Settler colonialism is an ongoing structure, not an event, upheld by institutions, policies and laws—many centuries old—that continue to shape renewable energy development on Native lands."

Through the course of this project, the goal was maintained alongside an openness to more efficient and appropriate alternatives to the initial research objective. Research was conducted with an eye for wind energy generation but with a willingness to acknowledge solar, geothermal or any other form of renewable energy as a superior option for the Fort Bidwell Reservation. For instance, it is known that geothermal energy has many advantages over solar and wind systems [9]. Such systems were briefly considered throughout this project.

The second area of research for this project is centered on the history of the Fort Bidwell Reservation as well as other tribes and their experiences in harnessing alternative energy on their land.

The native people of the Americas have a rich cultural heritage that runs parallel with a tragic and extensive history of colonialism. More recently, similar problems have arisen on their federally recognized lands: electricity inequality, educational gaps, and a generally lower standard of living [10–12]. It is true that there is a shared history, but it can also be said that due to cultural and geographic differences, each of these communities may tread different paths to an equitable future. Understanding the Fort Bidwell Indian Community is crucially important to this research when analyzing potential projects on their land. What may work for some communities may not work for all.

U.S. native tribes vary greatly in size. The largest U.S. tribe is the Navajo Nation, which, like the Cherokee, Choctaw, Chippewa, and Sioux, has a tribal membership of over 200,000 members [13]. On the other end of the spectrum, tribes like the Augustine Band of Cahuilla Indians have a total of 20 enrolled members [14]. The Fort Bidwell Indian Community is on the smaller side of the scale, according to the 2020 Census, 97 people reside on the Fort Bidwell Reservation [15]. The reservation has a slightly higher than national average employment rate of 62.3% and their median household income, $24,375, is less than one third the nation's [16].

A 2005 [17] study into the feasibility of installing a geothermal district heating system to provide low-cost heating by harnessing a geothermal well concluded that there is more than enough energy available from the FB-3 geothermal well to support the reservation and the entire town of Fort Bidwell. A second study in 2007 stated that drilling in a secondary location, FB-4, had begun in 2007 (Figure 1). This report stated that the FB-4 well had a temperature from 100 to 200 degrees Fahrenheit and, while the system would cost $1.5 million to install, it would have an annual energy savings of $124,300, yielding an estimated payback period of 12 years. In this report, consultant Dale Merrick went on to state that the Fort Bidwell Reservation could more than meet their own energy needs through geothermal means and that the reservation still had significant potential for developing both solar and wind energy to further bolster their future.

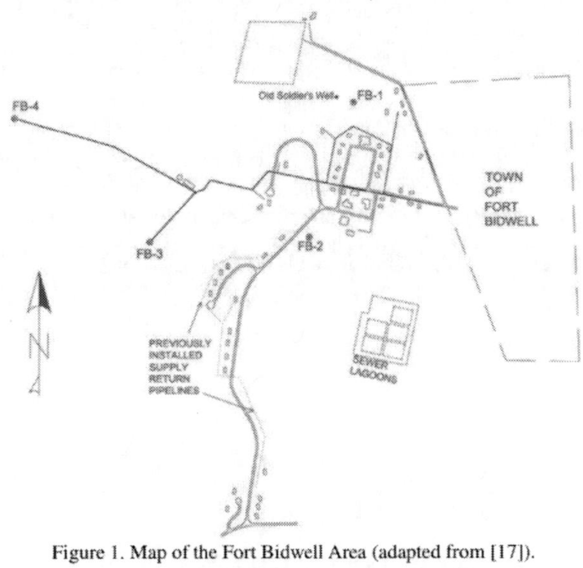

Figure 1. Map of the Fort Bidwell Area (adapted from [17]).

Of course, developing alternative energy on tribal lands is not a new idea. This being the case, there are examples of on-going projects to examine. One of the more enterprising tribes, The San Rincon Band of Luiseño Indians in the San Diego area, who may be most widely known as the owners of Harrah's Resort Southern California, have embarked on several energy projects to establish their own microgrid. Most recently, they received a grant for long-duration storage early in 2024 [18]. Prior to their latest investment into their energy sovereignty, they began work in 2020 to build a solar microgrid [19]. During the construction of this project, they issued a list of lessons learned that could be helpful to others attempting or planning similar endeavors. The first lesson was that microgrid projects can be extra complex; adding more components, such as meters, will ratchet up complexity.

Existing plans and information may be limited, rooftop solutions may not be feasible, and one may not be able to integrate existing energy assets. Furthermore, budget and time estimates should be ultra conservative, as costs will likely be higher than anticipated. Interconnection studies should be conducted as early as possible. A well-funded and resourced tribe like the San Rincon Band stressed that expert support is vital.

Not all native energy projects in California are as rosy. For example, the Campo Indian Reservation has been battling its neighbors in court to begin work on a vast wind farm to place 60 turbines on its land in San Diego. The project received approval in 2021, but, as of March 2024, the project was still tied up in litigation [20]. Other alternative energy projects have seen unobstructed success. For example, [21] reported that the Paskenta Band of Nomlaki Indians announced a plan to develop a large-scale solar and storage project on its reservation. The project was the recipient of a $32.75 million grant from the California Energy Commission under its Long Duration Storage Program and broke ground in April of 2024 [20]. The report went on to detail $31 million was awarded to Indian Energy LLC to develop a microgrid for the Viejas tribe of Kumeyaay Indians.

III. Data Selection and Acquisition

A. Data Source 1

The Global Wind Atlas [22] is a free, web-based application developed to help policymakers, planners, and investors identify high-wind areas for wind power generation. With this wind density mapping, areas with high potential for wind energy generation can be easily identified (see Figure 2).

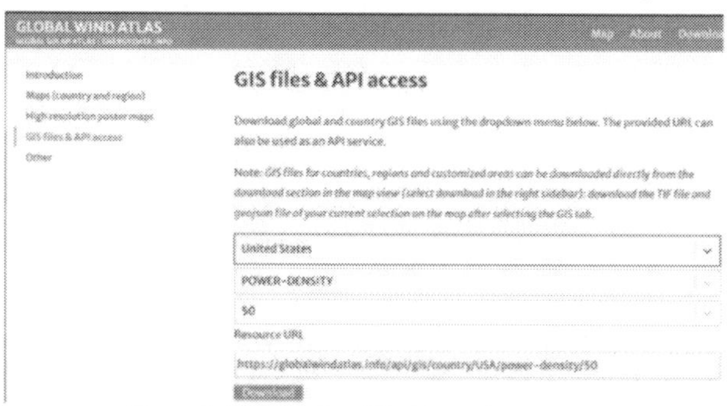

Figure 2. Downloading Wind Density Map [22].

B. Data Source 2

The California Energy Commission's federally recognized tribal lands map identifies tribal areas that fall within California's borders [23]. The data from the map is informed by the 2021 U.S. Census (see Figure 3).

C. Data Source 3

Building and geographical feature data came from OpenStreetMap (OSM) data [24] hosted by Esri, supported by Meta, and supplemented with additional data from Microsoft. The data is updated every month and aided in setting proper distances from existing infrastructure (see Figure 4).

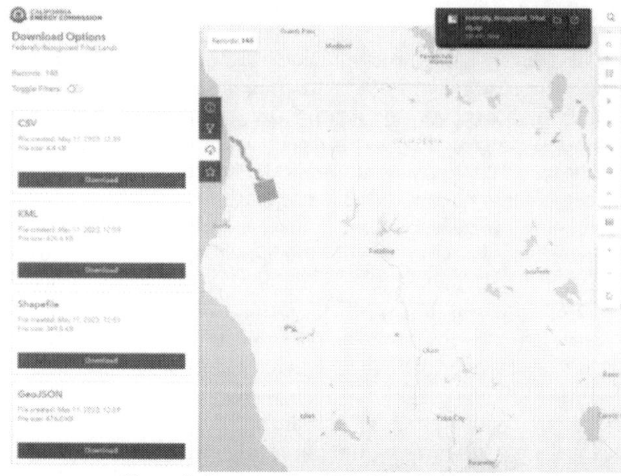

Figure 3. Downloading Tribal Land Map [23].

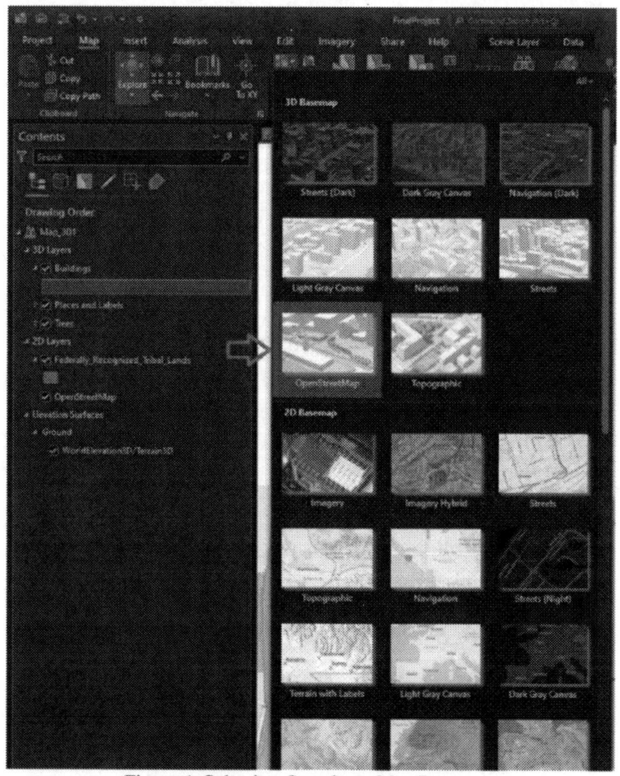

Figure 4. Selecting OpenStreetMap Basemap.

These data formed the basis of the analysis.

IV. System

The primary tool for this project is ArcGIS Pro 3.3.1. The aforementioned datasets were compiled onto a single map using ArcGIS Pro, and the buffer tool was deployed to visualize the areas where a wind farm would not be appropriate. The computer used was a Windows 10 machine with an Intel i7-7700K processor and a Nvidia RTX 2070 Super GPU.

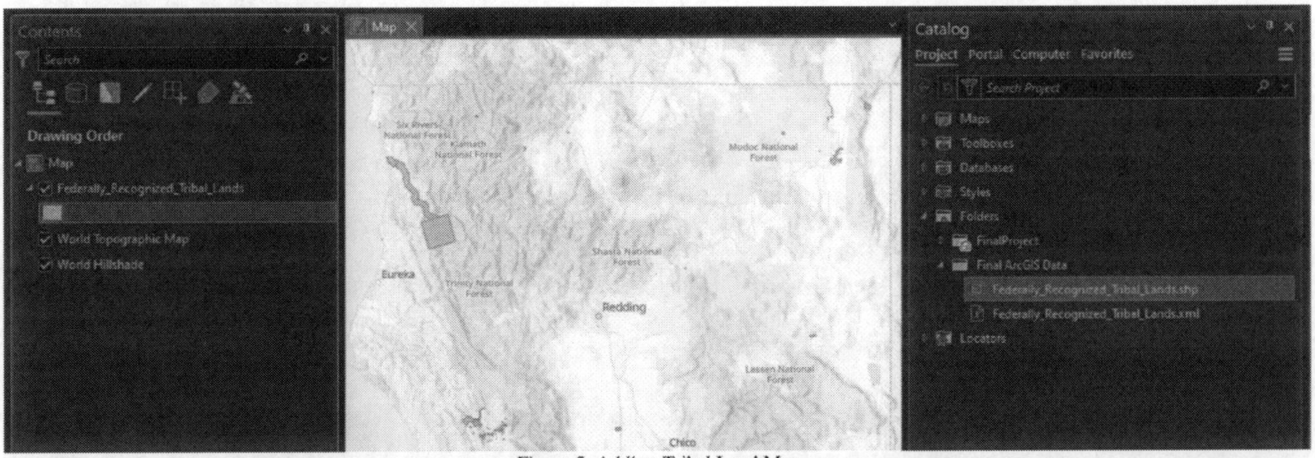

Figure 5. Adding Tribal Land Map.

V. METHODOLOGY

The first step of this project was adding the map of tribal land obtained from the California Energy Commission as a shapefile. The compressed folder was added to the project folder, extracted and then located in ArcGIS Pro in catalog view (see Figure 5).

Next, the wind density map from the Global Wind Atlas (as a TIFF image) was added to ArcGIS Pro. We selected Add Data under the Map tab. Figure 6 shows the results after it was processed.

OSM is a feature native to ArcGIS Pro under Basemap, this was used to identify roads, structures and water features. Figure 4 shows this process. These three steps form the basis of the project. Once these three elements were in place, more distinctive color schemes were chosen to differentiate the data. After that step, the main analysis tool was deployed. Buffer analysis was chosen for this project. This tool gives immediate visual information on which parts of the reservation would not be appropriate for a wind turbine. The buffer tool creates a visual radius around any feature selected by any distance specified. For this project, we used meters for the measurement as this was the usual unit of measurement in our cited references.

The first feature class processed using the buffer tool was Buildings containing residential-building information. The data of where these buildings were located was already present thanks to the basemap, OSM. A feature class named Buildings was created and the westernmost buildings were manually marked with yellow triangles (Figure 7).

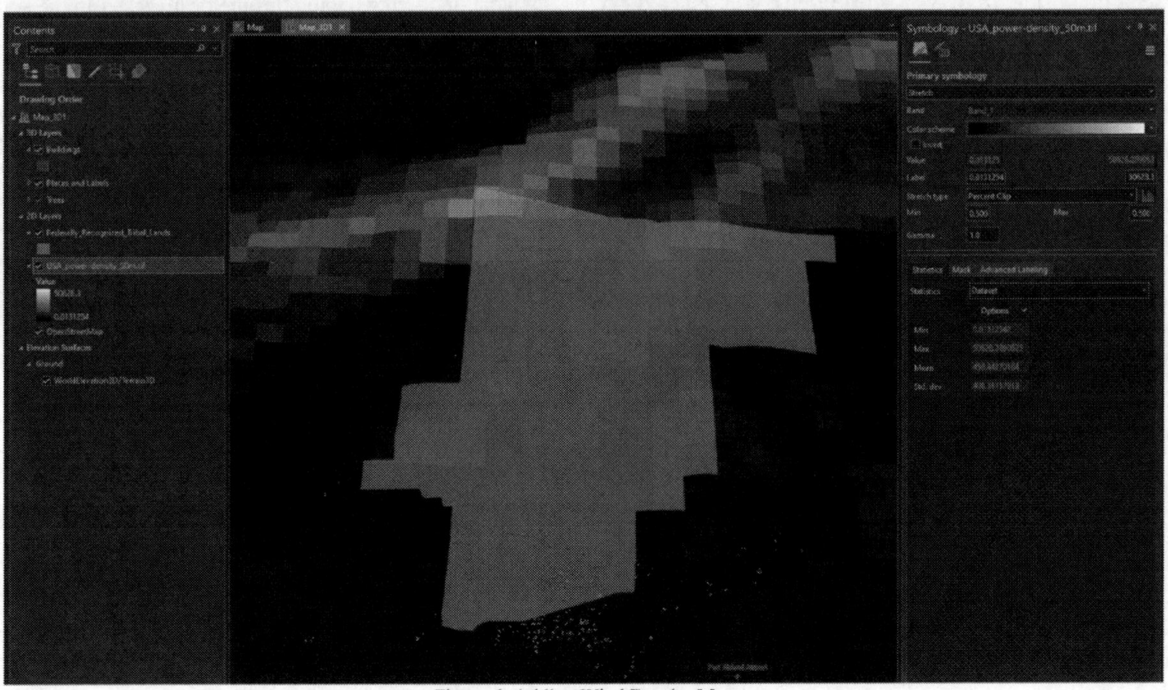

Figure 6. Adding Wind Density Map.

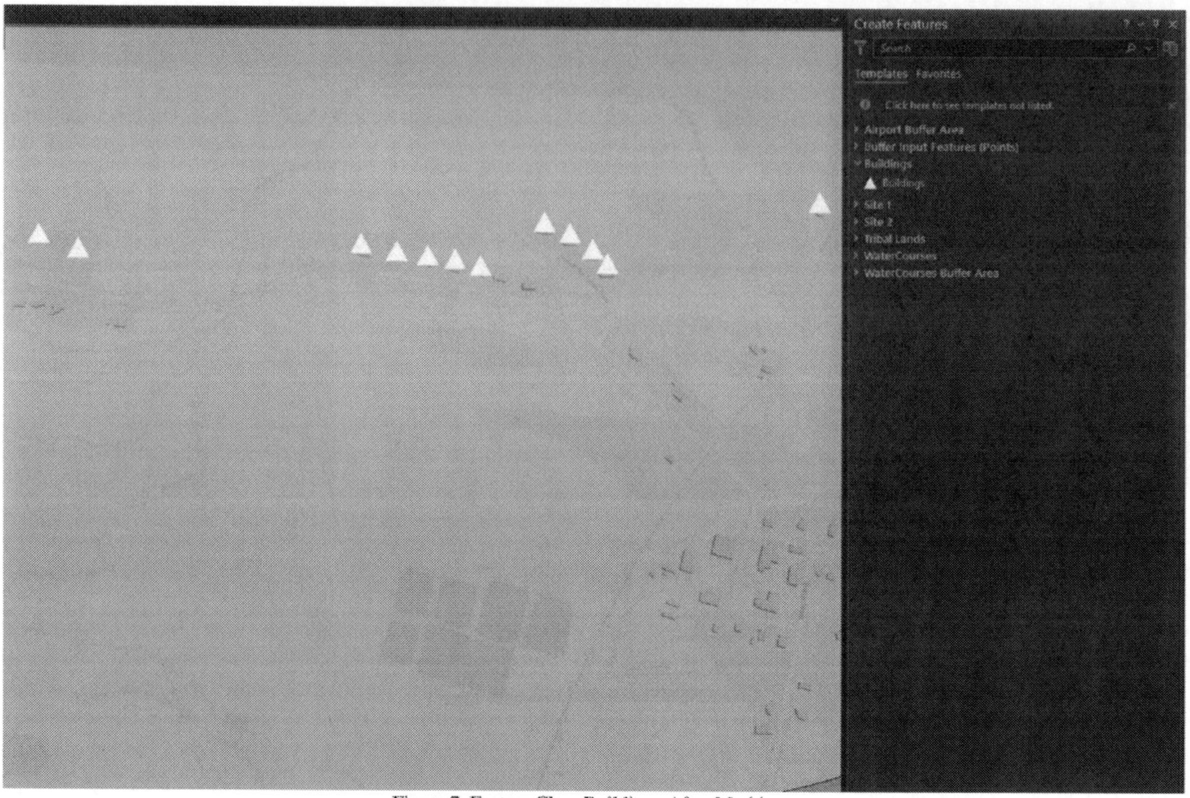

Figure 7. Feature Class Buildings After Marking.

The same process was followed for the local airport and all watercourses on the Fort Bidwell Reservation. These feature classes were named Airport and Watercourses respectively. Once all the restrictive factors were located and marked, the Buffer tool was then used. We selected the Buildings feature class, input the distance and chose the unit of measure (Figure 8).

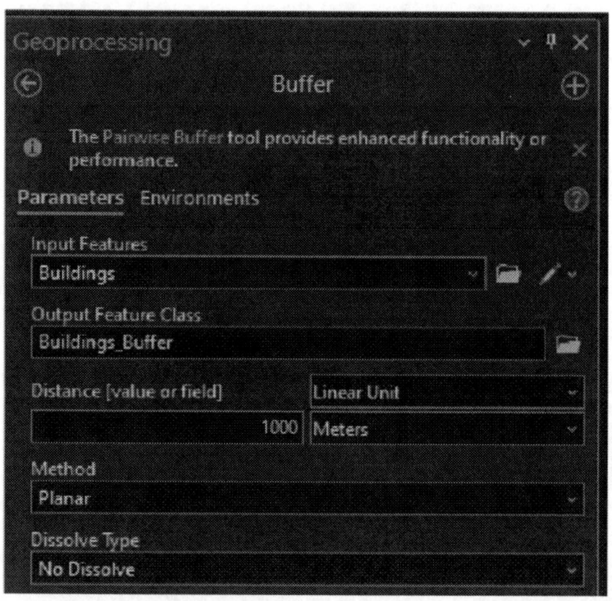

Figure 8. Buffer Tool Process for Buildings Feature Class.

VI. Results and Discussion

The analysis located two sites that could theoretically be capable of hosting a wind turbine farm: Site 1 in the southwest corner of the reservation and Site 2 along the western border of the reservation. Both sites, according to the area measurement tool, are over the 40-acre minimum requirement for a wind turbine operation. Site 1 is approximately 200 acres and Site 2 is approximately 90 acres. Figure 9 shows that the airport was a negligible obstruction. The residential areas are also far enough from the high-wind areas to matter little to site selection. The greatest restrictive factor was watercourses. The wind speeds on the reservation do not reach dangerous speeds, so this restriction was excluded from the analysis early in the project. Distance from roads was also eventually abandoned as a consideration because the roads near the two sites are dirt service roads that could be easily traversed.

VII. Conclusion and Future Work

The project resulted in useful data, but the sites need further scrutiny and expert input. We believe the project produced a solid starting point for energy-project evaluation on the reservation. The research portion highlighted the use of geothermal energy. Unfortunately, we were unable to locate any information on the success of these projects which would be of great value to any future energy project on the Fort Bidwell Reservation. Regardless of the success of those projects, wind or solar options would still be worth

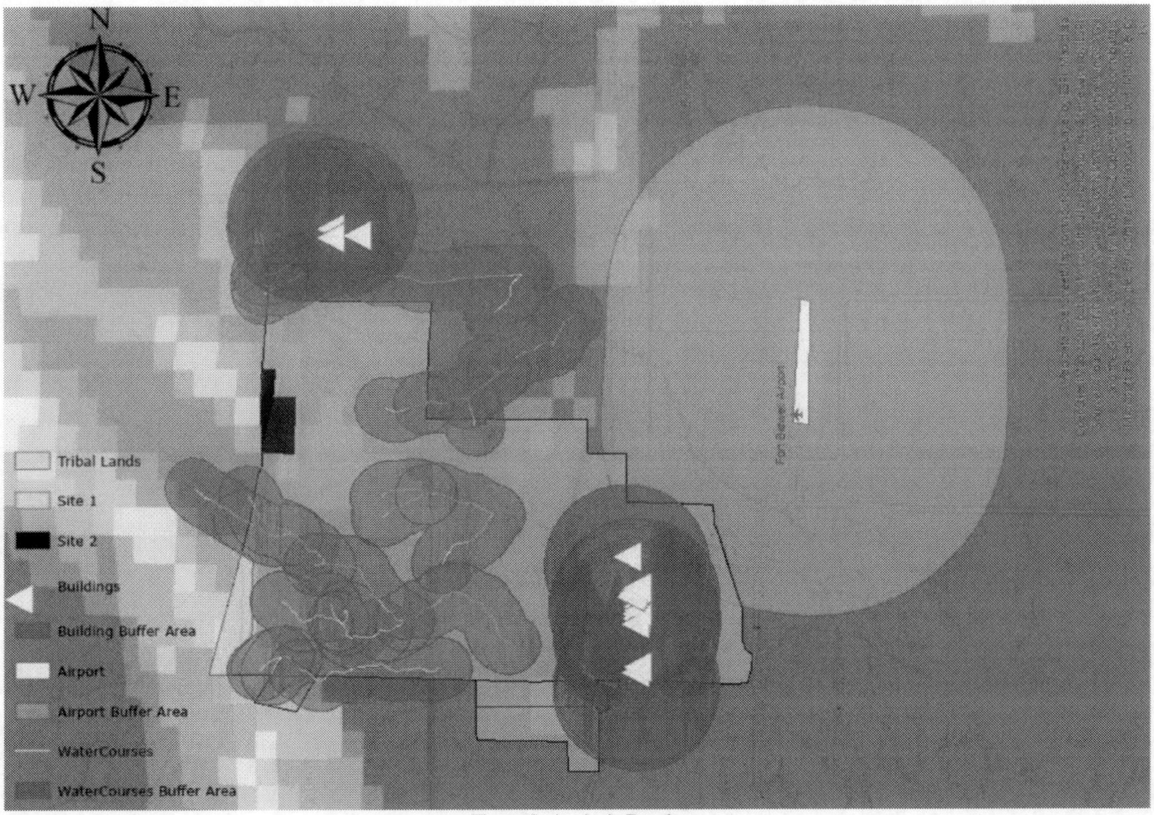

Figure 9. Analysis Results.

considering. Diversification of energy sources would help insulate the community from instability if any one source is disrupted by climate change, natural disaster, or technical challenges. We recommend for this particular area that anyone interested in beginning an energy project contact tribal leadership. This is a common hurdle for energy projects on tribal land. Furthermore, in some instances, appropriate land was determined, but projects ultimately failed when the tribal communities notified researchers that the land was of cultural importance. This made developing some projects infeasible on the grounds that the native communities would be uncooperative to defend their cultural heritage. In this specific case, according to research on the Fort Bidwell Reservation, geothermal energy had the potential to cover all energy needs of the community. If further research concludes that wind energy is not suitable, then it would be wise to survey the reservation for potential solar energy generation sites.

REFERENCES

[1] G. Rediske et al., "Wind power plant site selection: A systematic review," Renewable and Sustainable Energy Reviews, vol. 148, art. 111293, 2021. [Online]. Available from: https://doi.org/10.1016/j.rser.2021.111293

[2] C. Grosse and B. Mark, "Does renewable electricity promote indigenous sovereignty? Reviewing support, barriers, and recommendations for solar and wind energy development on native lands in the United States," Energy Research & Social Science, vol. 104, art. 103243, 2023. [Online]. Available from: https://doi.org/10.1016/j.erss.2023.103243

[3] P. Wolfe, "Settler colonialism and the elimination of the native," Journal of Genocide Research, vol. 8, no. 4, pp. 387–409, 1999. [Online]. Available from: https://doi.org/10.1080/14623520601056240

[4] K. Li, H. Bian, C. Liu, D. Zhang, Y. Yang, "Comparison of geothermal with solar and wind power generation systems." Renewable & Sustainable Energy Reviews, vol. 42, pp. 1464–1474, 2015. [Online]. Available from: https://doi.org/10.1016/j.rser.2014.10.049

[5] D. Merrick, "Renewable energy development on tribal lands," Energy.gov, 2007. [Online]. Available from: https://www.energy.gov/indianenergy/articles/project-reports-fort-bidwell-indian-community-2005-project. Accessed January 12, 2025.

[6] B. Edwards, "Northern California tribe sets path to energy sovereignty with large-scale solar and storage microgrid project," Tribal Business News, June 19, 2023. [Online]. Available from: https://tribalbusinessnews.com/sections/energy/14371-northern-california-tribe-sets-path-to-energy-sovereignty-with-large-scale-solar-and-storage-microgrid-project. Accessed January 12, 2025.

[7] M. Temp and R. Nikolewski, "Wind project in San Diego's backcountry runs into turbulence," San Diego Union-Tribune, March 22, 2024. [Online]. Available from: https://www.sandiegouniontribune.com/2024/03/22/wind-project-in-san-diegos-backcountry-runs-into-turbulence. Accessed January 12, 2025.

[8] Global Wind Atlas, "Introduction," 2015. [Online]. Available from: https://globalwindatlas.info/en/about/introduction. Accessed January 12, 2025.

[9] California Energy Commission, "Federal recognized tribal lands," 2023. [Online]. Available from: https://cecgis-caenergy.opendata.arcgis.com/datasets/21620043addf4c9da3540538671e47ba_0/explore?location=41.861547%2C-120.143269%2C13.21. Accessed January 12, 2025.

[10] OpenStreetMap, Esri. [Online]. Available from: https://openstreetmap.maps.arcgis.com

Security and Attacks on Federated Energy Forecasting

Jonas Sievers*, Krupali Kumbhani†, Thomas Blank*, Frank Simon*, Andreas Mauthe†

*Karlsruhe Institute of Technology (KIT), Institute for Data Processing and Electronics (IPE), Karlsruhe, Germany, e-mail: jonas.sievers@kit.edu
†University of Koblenz, Institute for Information Systems Research, Koblenz, Germany

***Abstract*—Accurate energy forecasting, including load, photovoltaic generation, and prosumption prediction, is essential for the efficient operation and strategic planning of modern energy systems. Federated Learning (FL) has emerged as a promising solution for training machine learning models on decentralized data, enabling high model accuracy while maintaining data privacy. However, the decentralized nature of FL also poses security challenges, including data poisoning and backdoor attacks that compromise the integrity and reliability of forecasting models. In this study, we present a comprehensive evaluation of various data poisoning and backdoor attacks within federated energy forecasting. Our analysis explores different data distributions, varying noise scales in data poisoning attacks, and targeted manipulation of specific time intervals to assess their impact on model performance. Further, we propose robust security mechanisms, such as increased cluster sizes, local retraining, and weighted aggregation. Our results show that while our attacks can increase the Mean Absolute Error by 93-261 %, our security measures can effectively mitigate the attacks, thereby improving the security and robustness of federated energy forecasting.**

Keywords- federated learning; poisoning attack; backdoor attack.

I. INTRODUCTION

The transition to sustainable energy systems is essential for addressing climate change and reducing the dependence on fossil fuels. As countries decarbonize their grids, accurate energy forecasting becomes critical to balance renewable energy supply and demand [1]. Reliable grid operation depends on accurate predictions of electric loads, photovoltaic (PV) generation, and prosumption patterns, especially as energy grids become more decentralized and complex [2].

Here, Federated Learning (FL) has been proposed for energy forecasting, enhancing model performance, data efficiency, and privacy. As shown in Figure 1, only model parameters are shared with a central server in FL, while local data remains private [3]. This approach minimizes the risk of exposing sensitive consumption patterns, which could otherwise be exploited to infer personal habits, posing privacy threats [4]. Additionally, clustering is applied in FL to group nodes with similar energy patterns, addressing challenges with non independent and identically distributed (non-iid) datasets.

While FL enhances privacy and security, challenges such as data poisoning and backdoor attacks persist. Data poisoning skews model performance by manipulating local data, while backdoor attacks insert hidden triggers into the model that only activate malicious behavior under specific conditions. While these vulnerabilities have been studied within the vision

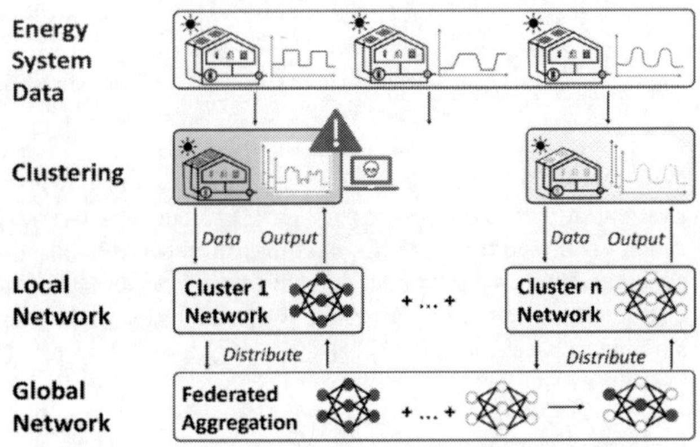

Figure 1. Clustered federated learning architecture with data poisoning.

and language domains [5], they are particularly concerning in energy systems, where forecasting errors can disrupt supply and demand balance, hindering renewable energy integration [6]. Addressing these risks is crucial to improve the robustness of FL in energy forecasting.

To address security challenges in FL-based energy forecasting applications, we investigate the effects of data poisoning and backdoor attacks. By manipulating data at varying scales and time periods, we assess model vulnerabilities, identifying weaknesses in the FL framework. We then propose defense strategies, including secure aggregation, local retraining, and clustering methods, to enhance the system's resilience against adversarial threats.

A. Related Work

To provide a comprehensive understanding of the current research and challenges in federated energy systems, we review related work on security measures and adversarial attacks. Additionally, we examine implementations of these attacks in the domains of computer vision and natural language processing. Selected publications are summarized in Table I.

McMahan et al. introduced FL in 2016 [3] as a method for decentralized model training across distributed devices, preserving data privacy by keeping data localized. While FL has been applied in the energy domain for applications like energy control [20]–[22], non-intrusive load monitoring [23]–[25], and energy theft detection [26], research on adversarial attacks within federated energy systems remains limited. Here, only

TABLE I
REVIEW OF SECURITY AND ATTACK LITERATURE IN FEDERATED ENERGY FORECASTING.

Ref	Year	Focus	Domain	Attack	Security
[7]	2021	Differential Privacy	Energy		✓
[8]	2023	Differential Privacy	Energy		✓
[9]	2022	Differential Privacy	Energy		✓
[10]	2024	Secure Aggregation	Energy		✓
[11]	2023	Secure Aggregation	Energy		✓
[12]	2023	Secure Aggregation	Energy		✓
[13]	2023	Secure Aggregation	Energy		✓
[14]	2022	Secure Aggregation	Energy		✓
[15]	2023	Secure Aggregation	Energy		✓
[16]	2023	Personalized FL	Energy		✓
[17]	2022	Model Poisoning	Energy	✓	
[18]	2020	Inference Attacks	Vision	✓	
[19]	2019	Poisoning Attacks	Language	✓	
This paper	2024	Attacks and Security	Energy	✓	✓

model poisoning has been analyzed [17], as most research focuses on security measures [7]–[16]. In contrast, adversarial attacks – including model poisoning, inference attacks, data poisoning, and backdoor attacks – have been extensively studied in domains like computer vision [18], [27], [28] and natural language processing [19], [29], [30], highlighting significant risks to FL models.

B. Paper Contribution and Organization

Vulnerabilities specific to federated energy forecasting have so far not been thoroughly investigated. To address this gap, we analyze adversarial attacks in federated energy forecasting and propose mitigation strategies. Since most research on FL attacks focuses on natural language processing and computer vision, applying those findings to energy prediction is challenging due to different data characteristics and dimensionality. Consequently, our main contributions are:

- We develop data poisoning and backdoor attacks customized for energy forecasting, evaluating their impact on model performance in FL systems using selected noise distributions (Uniform, Normal, Laplace, Building's) and targeting specific time intervals.
- We benchmark these attacks across different model architectures, including a Bidirectional Long-Short Term Memory Model (BiLSTM), a Soft-Gated LSTM (Soft-LSTM), and a Soft-Gated Dense Neural Network (Soft-Dense).
- We integrate security mechanisms such as secure aggregation, varying cluster sizes, and local retraining, to mitigate the effects of these attacks.
- Our findings show that data poisoning attacks significantly impact model performance, especially in small clusters, while backdoor attacks pose minor threats. By incorporating our proposed security measures, these adverse effects are mitigated, enhancing the security and robustness of FL-based energy forecasting systems.

The reminder of the paper is organized as follows: Section II introduces our methodology, while Section III outlines our experimental setup. Building on this, Section IV presents our results, Section V discusses our results and limitations, and Section VI provides our conclusion and future work.

II. Methodology

In this section, we provide a concise overview of our methodology, including federated energy systems, data poisoning, backdoor attacks and security measures.

A. Federated Energy Systems

In federated energy systems, a central server initializes global model weights w_0 for each cluster and distributes them to local devices (clients). Each client i trains a local model on its local dataset D_i and returns updated weights w_i. The server aggregates these weights using a selected aggregation method. One common approach is Average Aggregation, where the global model weights w_{global} are updated as:

$$w_{global} = \frac{1}{N}\sum_{i=1}^{N} w_i \tag{1}$$

where N is the number of clients within a cluster. This process is repeated over t federated training rounds.

B. Data Poisoning Attack in Federated Energy Systems

Data poisoning attacks threaten federated energy systems by compromising the integrity of the global FL model. Attackers manipulate the local datasets of specific clients, distorting energy forecasts and leading to inaccurate predictions that can affect operations like load balancing or grid management.

For a data point (x, y), with the input vector x and the target vector y, the poisoned input vector x' is defined as:

$$x' = x + \epsilon, \quad \epsilon \sim \mathcal{D}(\gamma) \tag{2}$$

Here, ϵ represents noise sampled from a distribution $\mathcal{D}$ with noise scale γ. Possible choices for $\mathcal{D}$ include Normal, Laplace, and Uniform distributions, or the distribution of the actual building measurements. To ensure proportionate noise injection across varying energy data scales, we normalize x to the range $[0, 1]$. During local training, attacked clients $j \in \mathcal{A}$ use the poisoned data x'_j, while benign clients $k \in \mathcal{B}$ use unmodified data. The weight update for benign clients is:

$$\mathbf{w}_{k,\text{new}} = \mathbf{w}_k - \eta \frac{\partial \mathcal{L}(\mathbf{w}_k, \mathbf{x}_k, y_k)}{\partial \mathbf{w}_k} \tag{3}$$

For attacked clients $j \in \mathcal{A}$, the weight update is:

$$\mathbf{w}'_{j,\text{new}} = \mathbf{w}_j - \eta \frac{\partial \mathcal{L}(\mathbf{w}_j, \mathbf{x}'_j, y_j)}{\partial \mathbf{w}_j} \tag{4}$$

Here, η is the learning rate, and $\frac{\partial \mathcal{L}}{\partial \mathbf{w}}$ denotes the gradient of the loss function $\mathcal{L}$ with respect to the model weights $\mathbf{w}$.

During the federated aggregation, the global model weights $\mathbf{w}'_{\text{global}}$ are updated by averaging:

$$\mathbf{w}'_{\text{global}} = \frac{1}{N}\left(\sum_{k \in \mathcal{B}} \mathbf{w}_{k,\text{new}} + \sum_{j \in \mathcal{A}} \mathbf{w}'_{j,\text{new}}\right) \tag{5}$$

where N is the total number of clients. The inclusion of poisoned weights $\mathbf{w}'_{j,\text{new}}$ can degrade global model performance by introducing biased patterns. Systematically testing different distributions $\mathcal{D}$ and noise scales γ allows us to evaluate the model's vulnerability to these attacks.

C. Backdoor Attack in Federated Energy Systems

Backdoor attacks in federated energy systems manipulate client data during specific hours, potentially causing the FL model to produce inaccurate forecasts during peak hours while maintaining normal performance at other times.

Therefore, attackers adjust the input data only during selected hours $H \subseteq 0, 1, \ldots, 23$. For each data point (t_j, x_j, y_j), where t_j denotes the hour, the modified input feature vector x'_j is defined as:

$$x'_j = \begin{cases} x_j + \delta, & \text{if } t_j \in H \\ x_j, & \text{otherwise} \end{cases} \quad (6)$$

Here, δ represents the backdoor trigger – a specific perturbation added only during the targeted hours H. The local training and federated aggregation remain the same as described in Subsection II-B. By adjusting δ and selecting specific hours H, attackers can fine-tune the severity of the backdoor attack and evaluate the model's performance.

D. Security Measures in Federated Energy Systems

To mitigate data poisoning and backdoor attacks in federated energy systems, we propose three security strategies: clustering, weighted aggregation, and local retraining.

Clustering reduces the impact of attacks by grouping clients with similar time series, restricting the extent of manipulation before a client is excluded from the cluster. Clients are grouped together when their time series E and F satisfy the similarity condition $d(E, F) \leq \tau$. A common similarity measure is Dynamic Time Warping (DTW), which minimizes the cumulative distance over all possible alignments (m_i, n_i):

$$d_{\text{DTW}}(E, F) = \sqrt{\min_{m_i, n_i} \left(\sum_{i=1}^{I} (e_{m_i} - f_{n_i})^2 \right)}, \quad (7)$$

Selecting an appropriate τ ensures that deviations from attacks remain within acceptable bounds.

Weighted Aggregation mitigates the attack effects by adjusting client contributions based on local model performance. Clients exhibiting degraded performance due to attacks receive lower weights, reducing their influence on the global model. Given a sets of benign clients $\mathcal{B}$ and attacked clients $\mathcal{A}$, the aggregation is performed as follows:

$$\mathbf{w}_{\text{global}} = \frac{\sum_{k \in \mathcal{B}} \alpha_k \cdot \mathbf{w}_k + \sum_{j \in \mathcal{A}} \alpha'_j \cdot \mathbf{w}'_j}{\sum_{k \in \mathcal{B}} \alpha_k + \sum_{j \in \mathcal{A}} \alpha'_j} \quad (8)$$

where α_k is the weight for benign clients, and α'_j represents the weight for attacked clients, typically $\alpha'_j \ll \alpha_k$.

Local Retraining allows benign clients to adapt the global model to their local data, reducing the impact of poisoned global weights w'_{global}. After receiving $\mathbf{w}'_{\text{global}}$, benign clients refine their models:

$$\mathbf{w}_{k,\text{retrained}} = \mathbf{w}_{\text{global}} - \eta \frac{\partial \mathcal{L}(\mathbf{w}_{\text{global}}, \mathbf{x}_k, y_k)}{\partial \mathbf{w}_{global}} \quad (9)$$

where η is the learning rate, $\mathcal{L}$ is the loss function, and $(\mathbf{x}_k, y_k)$ represents the local dataset. This fine-tuning mitigates attack influence and enhances model robustness.

Together, these strategies enhance the security and reliability of federated models in energy systems by effectively countering adversarial attacks.

III. Experimental Setup

Building on our methodology, we describe our experimental setup, including data analysis and federated energy forecasting.

A. Data Analysis

We utilize the Ausgrid dataset [31], which provides half-hourly smart meter readings of electrical load and PV output in kW from 300 residential buildings in Australia between 2010 and 2013. An example of load and PV patterns for Building 11 is shown Figure 2.

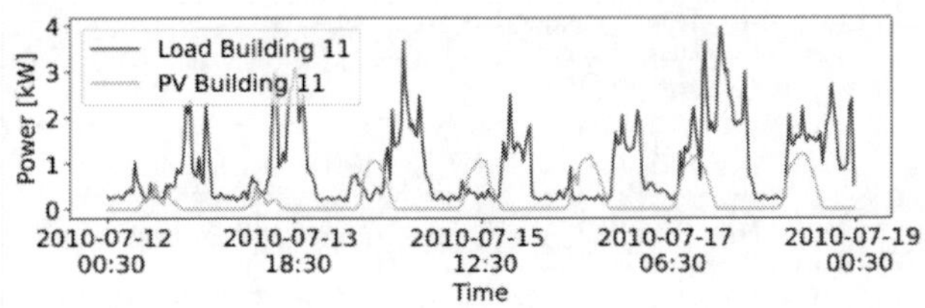

Figure 2. Load and PV patterns of Building 11.

We extend the dataset by calculating prosumption (load - PV). For computational efficiency, our analysis focuses on a randomly selected subset of the first 30 households. To obtain our forecasting dataset, we use 70% for training, 20% for validation, and 10% for testing.

B. Federated Energy Forecasting

Our FL architecture consists of 3 training rounds, as additional rounds did not yield further improvements. In each round, clients update their local models and send them to a central server for global aggregation. We use K-Means clustering with DTW to group clients with similar energy patterns into 10 clusters. For testing adversarial attacks, we focus on a cluster containing 2 clients (buildings 16 and 24), simulating distributed training on a single machine.

To implement federated energy forecasting, we employ a BiLSTM network and two Mixture of Experts (MoE) models within the FL framework. The MoE architecture enhances the model's ability to learn complex patterns by dynamically selecting and weighting outputs from multiple specialized submodels (experts) based on the input sequences. Specifically, the Soft-Dense model includes an expert layer with four experts (each with eight units), followed by two Dense layers (16 units each), a Dropout layer (rate 0.2), and a Flatten layer.

The Soft-LSTM model comprises an expert layer (four experts, eight units each), a bidirectional LSTM layer (four units), a Dropout layer (rate 0.2), and a Flatten layer. The BiLSTM network captures temporal dependencies in energy data through bidirectional processing, using two layers of eight LSTM cells each. All models use a batch size of 16. Further architectural details are provided in [4].

To evaluate robustness against adversarial attacks, we implement data poisoning and backdoor scenarios. In the data poisoning attack, noise is injected into the data from various distributions. For Normal and Laplace distributions, the mean is 0, and the standard deviation varies from 0 to 1. For the Uniform distribution, bounds range between $[-1, 0]$ and $[0, 1]$. Building-specific distributions are derived based on skewness, kurtosis, and mean, with standard deviations varying between 0 and 1. In the backdoor attack, we manipulate four specific half-hour time steps during hours 0 and 1, setting load and prosumption values to zero and PV values to one.

As security measures, we expand the cluster size to 4 buildings and implement a local retraining step with 100 epochs, incorporating early stopping if the validation loss does not improve over 10 consecutive epochs. Weighted aggregation is not employed in this experiment, as clustering and retraining already mitigate the effects. For performance evaluation, we calculate the Mean Absolute Error (MAE):

$$\text{MAE} = \frac{1}{n}\sum_{i=1}^{n} |\hat{y}_i - y_i| \tag{10}$$

This metric quantifies the average absolute difference between the predicted values $\hat{y}_i$ and actual values y_i.

IV. Results

In this section, we present our results for data poisoning, backdoor attacks, and security measures. Within each attack scenario, the *poisoned* model is trained with manipulated data, while the *unmodified* model is only indirectly affected through FL. To consider statistical variations, each model is trained 3 times per scenario. If not stated otherwise, the indicated metrics are averaged over all buildings, clusters, or training rounds and the results are achieved on the test dataset.

A. Data Poisoning in Federated Energy Forecasting

Within the data poisoning attack, noise patterns are introduced to the training data using four distributions: Normal, Laplace, Uniform, and building-specific. Due to space constraints, we report results only for the Uniform distribution, which had the most significant effects.

Figure 3 shows the MAE for both poisoned models (top row) and unmodified models (bottom row) with increasing noise scales over load, PV, and prosumption forecasting. The dashed lines represent the baseline performance of unmodified models within local learning, to evaluate whether the attacked FL architecture still provides performance benefits.

The results indicate a substantial increase in MAE with rising noise scale for all poisoned models. Specifically, the Soft-Dense model exhibits increases in MAE of 321 % for load, 718 % for PV, and 345 % for prosumption forecasting. Similar trends are observed in the BiLSTM and Soft-LSTM models. It is worth noting that as the MAE values are small, minor increases already result in large percentage changes. The unmodified models, indirectly affected through FL, also showes significant degradation, with MAE increases of up to 96% for load, 93% for PV, and 261% for prosumption forecasting in the Soft-Dense model.

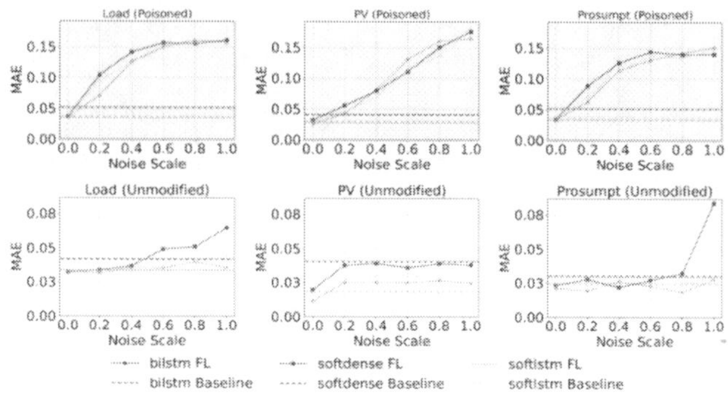

Figure 3. Performance comparison between poisoned and unmodified models.

Without poisoning, FL generally outperforms local learning. However, when subject to attack, the unmodified Soft-Dense model's performance droped below the local learning baseline for load forecasting at a noise scale of 0.6 and for prosumption prediction at 1.0. In contrast, PV forecasting performance remained close to the local baseline across all noise scales. Detailed results for the unmodified Soft-Dense model are provided in Table II.

TABLE II
MODEL PERFORMANCE OF THE UNMODIFIED SOFT-DENSE MODEL FOR LOCAL LEARNING (LL) AND FL WITH DIFFERENT NOISE SCALES (N), WHERE N0.2 CORRESPONDS TO A NOISE SCALE OF 0.2.

Noise	Load		PV		Prosumption	
	MAE	STD	MAE	STD	MAE	STD
LL	0.0417	±0.0051	0.0405	±0.0055	0.0302	±0.0068
N0	0.0324	±0.0196	0.0196	±0.0164	0.0234	±0.0140
N0.2	0.0338	±0.0047	0.0377	±0.0248	0.0278	±0.0023
N0.4	0.0369	±0.0016	0.0391	±0.0258	0.0219	±0.0017
N0.6	0.0492	±0.0007	0.0360	±0.0265	0.0271	±0.0029
N0.8	0.0511	±0.0028	0.0389	±0.0268	0.0323	±0.0079
N1	0.0649	±0.0255	0.0379	±0.0261	0.0836	±0.0012

B. Backdoor Attack in Federated Energy Forecasting

In the backdoor attack scenario, data is selectively modified for specific hours, using the date as the trigger. Figure 4 shows the MAE for each hour of the day for both poisoned models (top row) and unmodified models (bottom row). Within each subplot, solid lines represent the performance of the FL architectures affected by the attack, while dashed lines indicate the baseline performance without any attack. The hourly MAE naturally fluctuates due to inherent volatility variations.

For the Soft-Dense model with the backdoor attack, the MAE increased significantly only in prosumption forecasting, rising

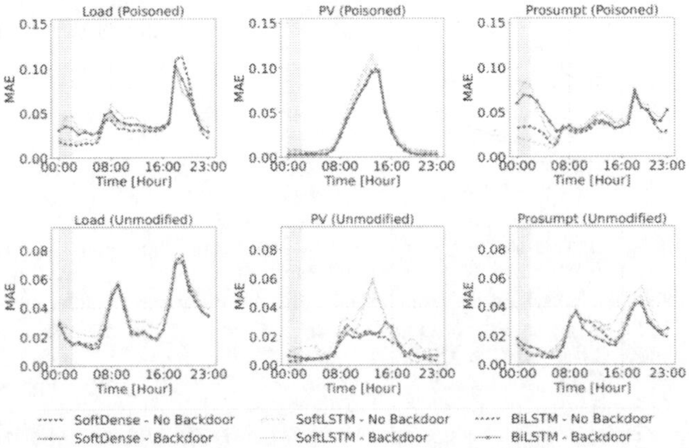

Figure 4. Model performance comparison with and without backdoor attacks.

by 0.0560 (273%). Changes in load (0.0142, 68%) and PV (0.0022, 33%) forecasting remain within the standard deviation. The unmodified models show minimal changes, with slight increases in load (0.0083, 36%) and prosumption (0.0001, 1%), and a decrease in PV (-0.0051, -49%), all within the standard deviation. Detailed MAE values of the Soft-Dense model for the attacked hours are provided in Table III, comparing the backdoor model to the baseline.

TABLE III
PERFORMANCE METRICS OF THE FORECASTING MODELS WITH AND WITHOUT BACKDOOR ATTACKS.

Scenario	Load		PV		Prosumpt.	
	MAE	STD	MAE	STD	MAE	STD
Pois. Back.	0.0351	0.0167	0.0089	0.0064	0.0765	0.0164
Pois. noBack.	0.0209	0.0110	0.0067	0.0017	0.0205	0.0105
Pois. Diff.	0.0142	0.0057	0.0022	0.0047	0.0560	0.0059
Un. Back.	0.0312	0.0073	0.0054	0.0023	0.0127	0.0039
Un. noBack.	0.0229	0.0063	0.0105	0.0060	0.0126	0.0025
Un. Diff.	0.0083	0.0010	-0.0051	-0.0037	0.0001	0.0014

C. Security in Federated Energy Forecasting

We mitigate attacks by increasing the cluster size, reducing the influence of compromised buildings on the aggregation model, while local retraining refines model parameters using unmodified data. Due to space constraints, Figure 5 illustrates that the MAE remains stable for all unmodified models and noise scales, indicating that this combination effectively mitigates the attacks.

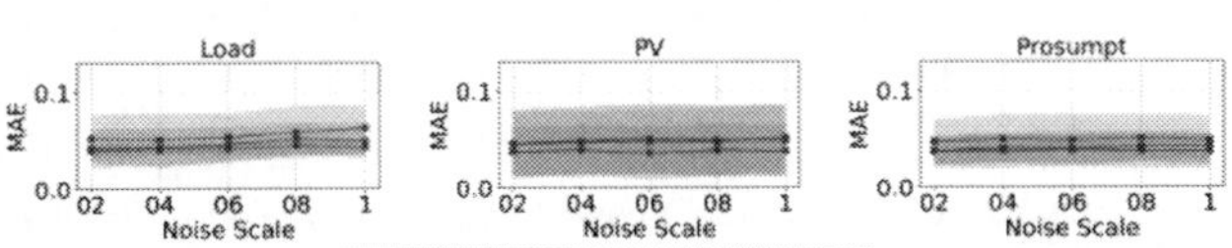

Figure 5. Impact of security measures on the unmodified forecasting models.

While these measures are effective, incorporating weighted average aggregation could further improve resilience, by weighting models based on their performance, thus reducing the contribution of compromised models.

V. Discussion and Limitations

Our data poisoning attacks substantially increased the MAE across all models, primarily due to the limited cluster sizes that increase the impact of compromised data. Load forecasting, due to its inherently stochastic nature, experienced greater performance degradation, whereas PV prediction maintained more stable performance, benefiting from its more deterministic patterns. Conversely, backdoor attacks had a minimal overall effect, as their targeted manipulations were limited to specific forecasting times, thereby reducing their influence on the aggregated model. The implemented security measures – namely, increasing cluster size and applying local retraining – effectively mitigate these attacks by reducing the impact of poisoned data. In addition, incorporating weighted average aggregation could further enhance resilience by reducing the contribution of compromised models during federated aggregation.

This study is limited by its focus on only few buildings, which may affect the generalizability of our findings. Incorporating a broader range of benchmark models and datasets could further validate the robustness of the proposed defense mechanisms. Further, we use simple noise sampling methods for data poisoning, which may not fully capture the complexity of more advanced attacks.

VI. Conclusion and Future Work

In this paper, we comprehensively analyzed security vulnerabilities in federated energy forecasting, focusing on the impacts of data poisoning and backdoor attacks. Our findings demonstrate that data poisoning poses a significant threat to forecasting accuracy, with MAE increasing by 93-261 %, especially within smaller clusters. Conversely, backdoor attacks show a limited impact on model performance. By incorporating defense mechanisms such as increased cluster sizes and local retraining, we effectively enhanced the resilience of federated learning models, mitigating the adversarial risks and preserving model integrity. Future work could explore the use of Generative Adversarial Networks for sophisticated noise generation during training.

References

[1] F. Plaum, R. Ahmadiahangar, A. Rosin, and J. Kilter, "Aggregated demand-side energy flexibility: A comprehensive review on characterization, forecasting and market prospects", *Energy Reports*, vol. 8, pp. 9344–9362, 2022, ISSN: 2352-4847. DOI: https://doi.org/10.1016/j.egyr.2022.07.038.

[2] U. Das *et al.*, "Forecasting of photovoltaic power generation and model optimization: A review", *Renewable Sustainable Energy Reviews*, vol. 81, pp. 912–928, 2018. DOI: 10.1016/J.RSER.2017.08.017.

[3] H. B. McMahan, E. Moore, D. Ramage, S. Hampson, and B. A. y Arcas, "Communication-efficient learning of deep networks from decentralized data", in *International Conference on Artificial Intelligence and Statistics*, 2016.

[4] J. Sievers, T. Blank, and F. Simon, "Advancing Accuracy in Energy Forecasting using Mixture-of-Experts and Federated Learning", in *Proceedings of the 15th ACM International Conference on Future and Sustainable Energy Systems*, ser. e-Energy '24, Singapore, Singapore: Association for Computing Machinery, 2024, pp. 65–83, ISBN: 9798400704802. DOI: 10.1145/3632775.3661945.

[5] J. Zhang *et al.*, "Security and Privacy Threats to Federated Learning: Issues, Methods, and Challenges", *Sec. and Commun. Netw.*, vol. 2022, 2022, ISSN: 1939-0114. DOI: 10.1155/2022/2886795.

[6] J. Jasiūnas, P. D. Lund, and J. Mikkola, "Energy system resilience – A review", *Renewable and Sustainable Energy Reviews*, vol. 150, p. 111 476, 2021, ISSN: 1364-0321. DOI: https://doi.org/10.1016/j.rser.2021.111476.

[7] Y. Zhao *et al.*, "A Differential Privacy-enhanced Federated Learning Method for Short-Term Household Load Forecasting in Smart Grid", Institute of Electrical and Electronics Engineers Inc., 2021, pp. 1399–1404, ISBN: 9781665409506. DOI: 10.1109/ICCC54389.2021.9674514.

[8] X. Qu *et al.*, "Personalized Federated Learning for Heterogeneous Residential Load Forecasting", *Big Data Mining and Analytics*, vol. 6, pp. 421–432, 4 2023, ISSN: 20960654. DOI: 10.26599/BDMA.2022.9020043.

[9] M. A. Husnoo *et al.*, "A Secure Federated Learning Framework for Residential Short Term Load Forecasting", 2022.

[10] Y. Dong *et al.*, "Privacy-Preserving Distributed Learning for Residential Short-Term Load Forecasting", *IEEE Internet of Things Journal*, 2024, ISSN: 23274662. DOI: 10.1109/JIOT.2024.3362587.

[11] J. Li, H. Li, R. Wang, Y. Guo, and S. Wu, "Fed-SAD:A secure aggregation federated learning method for distributed load forecasting", 2023. DOI: 10.22541/au.169028986.64063960/v1.

[12] M. A. Husnoo *et al.*, "FedDiSC: A computation-efficient federated learning framework for power systems disturbance and cyber attack discrimination", *Energy and AI*, vol. 14, 2023, ISSN: 26665468. DOI: 10.1016/j.egyai.2023.100271.

[13] Y. Liu, Z. Dong, B. Liu, Y. Xu, and Z. Ding, "FedForecast: A federated learning framework for short-term probabilistic individual load forecasting in smart grid", *International Journal of Electrical Power and Energy Systems*, vol. 152, 2023, ISSN: 01420615. DOI: 10.1016/j.ijepes.2023.109172.

[14] M. M. Badr *et al.*, "Privacy-Preserving Federated-Learning-Based Net-Energy Forecasting", vol. 2022-March, Institute of Electrical and Electronics Engineers Inc., 2022, pp. 133–139, ISBN: 9781665406529. DOI: 10.1109/SoutheastCon48659.2022.9764093.

[15] H. U. Manzoor, A. R. Khan, T. Sher, W. Ahmad, and A. Zoha, "Defending Federated Learning from Backdoor Attacks: Anomaly-Aware FedAVG with Layer-Based Aggregation", Institute of Electrical and Electronics Engineers Inc., 2023, ISBN: 9781665464833. DOI: 10.1109/PIMRC56721.2023.10293950.

[16] F. Widmer, S. Nowak, B. Bowler, P. Huber, and A. Papaemmanouil, "Data-driven comparison of federated learning and model personalization for electric load forecasting", *Energy and AI*, vol. 14, 2023, ISSN: 26665468. DOI: 10.1016/j.egyai.2023.100253.

[17] N. B. S. Qureshi, D. H. Kim, J. Lee, and E. K. Lee, "Poisoning Attacks against Federated Learning in Load Forecasting of Smart Energy", Institute of Electrical and Electronics Engineers Inc., 2022, ISBN: 9781665406017. DOI: 10.1109/NOMS54207.2022.9789884.

[18] X. Luo and X. Zhu, "Exploiting defenses against gan-based feature inference attacks in federated learning", *CoRR*, vol. abs/2004.12571, 2020.

[19] J. Zhang, J. Chen, D. Wu, B. Chen, and S. Yu, "Poisoning attack in federated learning using generative adversarial nets", Institute of Electrical and Electronics Engineers Inc., 2019, pp. 374–380, ISBN: 9781728127767. DOI: 10.1109/TrustCom/BigDataSE.2019.00057.

[20] S. Lee, L. Xie, and D.-H. Choi, "Privacy-Preserving Energy Management of a Shared Energy Storage System for Smart Buildings: A Federated Deep Reinforcement Learning Approach", *Sensors*, vol. 21, no. 14, 2021, ISSN: 1424-8220. DOI: 10.3390/s21144898.

[21] S. Lee and D.-H. Choi, "Federated Reinforcement Learning for Energy Management of Multiple Smart Homes With Distributed Energy Resources", *IEEE Transactions on Industrial Informatics*, vol. 18, no. 1, pp. 488–497, 2022. DOI: 10.1109/TII.2020.3035451.

[22] F. Rezazadeh and N. Bartzoudis, "A Federated DRL Approach for Smart Micro-Grid Energy Control with Distributed Energy Resources", 2022. DOI: 10.1109/CAMAD55695.2022.9966919.

[23] A. Giuseppi, S. Manfredi, D. Menegatti, A. Pietrabissa, and C. Poli, "Decentralized Federated Learning for Nonintrusive Load Monitoring in Smart Energy Communities", in *2022 30th Mediterranean Conference on Control and Automation (MED)*, 2022, pp. 312–317. DOI: 10.1109/MED54222.2022.9837291.

[24] Y. Wang, I. L. Bennani, X. Liu, M. Sun, and Y. Zhou, "Electricity Consumer Characteristics Identification: A Federated Learning Approach", *IEEE Transactions on Smart Grid*, vol. 12, no. 4, pp. 3637–3647, 2021. DOI: 10.1109/TSG.2021.3066577.

[25] Y. He, F. Luo, G. Ranzi, and W. Kong, "Short-Term Residential Load Forecasting Based on Federated Learning and Load Clustering", in *2021 IEEE International Conference on Communications, Control, and Computing Technologies for Smart Grids (SmartGridComm)*, 2021, pp. 77–82. DOI: 10.1109/SmartGridComm51999.2021.9632314.

[26] M. M. Ashraf *et al.*, "FedDP: A Privacy-Protecting Theft Detection Scheme in Smart Grids Using Federated Learning", *Energies*, vol. 15, no. 17, 2022, ISSN: 1996-1073. DOI: 10.3390/en15176241.

[27] R. Mayerhofer and R. Mayer, "Poisoning Attacks against Feature-Based Image Classification", in *Proceedings of the Twelfth ACM Conference on Data and Application Security and Privacy*, ser. CODASPY '22, Baltimore, MD, USA: Association for Computing Machinery, 2022, pp. 358–360, ISBN: 9781450392204. DOI: 10.1145/3508398.3519363.

[28] Z. Xiang, D. J. Miller, and G. Kesidis, "A Benchmark Study Of Backdoor Data Poisoning Defenses For Deep Neural Network Classifiers And A Novel Defense", in *2019 IEEE 29th International Workshop on Machine Learning for Signal Processing (MLSP)*, 2019, pp. 1–6. DOI: 10.1109/MLSP.2019.8918908.

[29] S. Zhai *et al.*, "Text-to-Image Diffusion Models can be Easily Backdoored through Multimodal Data Poisoning", in *Proceedings of the 31st ACM International Conference on Multimedia*, ser. MM '23, Ottawa ON, Canada: Association for Computing Machinery, 2023, pp. 1577–1587, ISBN: 9798400701085. DOI: 10.1145/3581783.3612108.

[30] A. Wan, E. Wallace, S. Shen, and D. Klein, "Poisoning Language Models During Instruction Tuning", in *Proceedings of the 40th International Conference on Machine Learning*, A. Krause *et al.*, Eds., ser. Proceedings of Machine Learning Research, vol. 202, PMLR, 2023, pp. 35 413–35 425.

[31] Elizabeth L. Ratnam, Steven R. Weller, Christopher M. Kellett and Alan T. Murray, "Residential load and rooftop PV generation: an Australian distribution network dataset", *International Journal of Sustainable Energy*, vol. 36, no. 8, pp. 787–806, 2017. DOI: 10.1080/14786451.2015.1100196.

SECNIR: A Multi-Year Electricity Consumption Dataset of 881 French Companies in Islands and Overseas Regions

Pierre-Alexis Chevreuil
Pierre Costini
Eco CO2 Nice, France
e-mail: pierrealexis.chevreuil@gmail.com
e-mail: pierre.costini@centraliens.net

Antoine Tavant
Cédric Simi
Eco CO2, Lyon, France
e-mail: antoine.tavant@m4x.org
e-mail: cedric.simi@proton.me

Joris Costes
Eco CO2 and *Univ. Paris-Saclay, ENS Paris-Saclay, CNRS, Centre Borelli,*
Gif-sur-Yvette, France
e-mail: joris.costes@ens-paris-saclay.fr

Abstract—**This paper presents the Survey of Electricity Consumption in Non-Interconnected Regions (SECNIR), a multi-year electricity consumption dataset comprising data from 881 companies and local authorities in French overseas regions. To the best of our knowledge, it is the first large-scale electricity consumption open dataset for professionals in islands and overseas regions at the level of individual electricity meters. The dataset includes electricity consumption, weather, and indoor comfort data from a variety of professionals in six different French islands and overseas regions: Reunion Island, French Guiana, Martinique, Guadeloupe, Corsica, and Mayotte. We analyze how local climate influences electricity consumption, in particular in tropical regions. Additionally, we provide insights into the most power-intensive sectors and examine the variability of energy consumption within specific professional sectors.**

Keywords-Energy; Electricity; Smart meters; Non-Interconnected Regions; Thermosensitivity; Dataset.

I. Introduction

With the increasing share of renewable energies in the worldwide electricity mix, determining where and when electricity is consumed becomes crucial for the stability of electrical grids. The deployment of smart power meters has enabled large-scale studies on individuals [1]–[3] and companies [4], [5]. However, these studies almost always focus on parts of the world connected to large electrical grids and in mild climates.

To the best of our knowledge, no large-scale open dataset with individual meter consumption data exists for Non-Interconnected Regions (NIR). This is notably true for tropical regions, which are particularly interesting to study because of their different thermosensitivity compared to territories with milder climates.

TABLE I. Comparison of the mean direct CO_2 emissions per electrical kWh and mean electrical production cost in different French territories [6]–[10].

Region	Mean direct emissions $[gCO_2 \cdot kWh^{-1}]$	Mean production cost $[€ \cdot MWh^{-1}]$
Mainland France	32.4 (2023)	72-90 (2020)
Guadeloupe	703 (2019)	351 (2022)
Martinique	575 (2020)	323 (2022)
Reunion Island	732 (2021)	267 (2022)
Corsica	564 (2020)	316 (2022)
French Guiana	468 (2019)	274 (2022)

Obtaining data in such regions is generally challenging, as grid operators tend to deploy fewer smart meters compared to the mainland. For example, in continental France, over 90% of consumers have smart power meters, such as the Linky meters deployed by Enedis [11], while much fewer are installed in French overseas territories. Except for Corsica, very few electricity consumption data are available for these regions. Moreover, these territories heavily rely on fossil energies for electricity generation, leading to significant financial and environmental costs, as shown in Table I. In mainland France, the first cost figure represents the direct production cost, excluding network, import, and export costs, while the second corresponds to the annualized cost. Therefore, understanding how professionals use electricity in these regions and helping them to reduce their consumption is essential. Raising awareness about climate issues and promoting energy savings is an effective way to achieve this goal. Energy savings certificate programs are a French policy measure designed for this purpose. Under this scheme, energy suppliers must either achieve energy savings themselves or purchase certificates from other parties to meet their targets. These programs, defined by ministerial decrees, focus on providing information, training, funding and innovation to manage energy demand and alleviate household energy insecurity.

The data presented in this paper, collected through the SEIZE program managed by Eco CO2, has been used to create the anonymized, weekly-resampled Survey of Electricity Consumption in Non-Interconnected Regions (SECNIR) dataset. Figure 1 illustrates the locations of the 881 sensors.

This paper is organized as follows: Section II describes the SEIZE program and data collection methodology. Section III details the dataset structure and processing techniques. In Section IV, the analysis and visualization of the dataset is presented, followed by a discussion of its limitations. Finally, Section V concludes the paper and suggests directions for future research.

II. Program Description

This section describes the deployment and methodology of the SEIZE program, which aims to collect and analyze electricity consumption data from various French overseas regions.

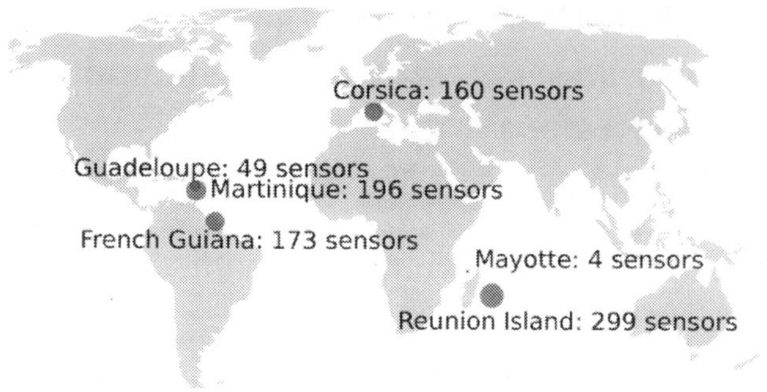

Figure 1. Number of electrical sensors per region.

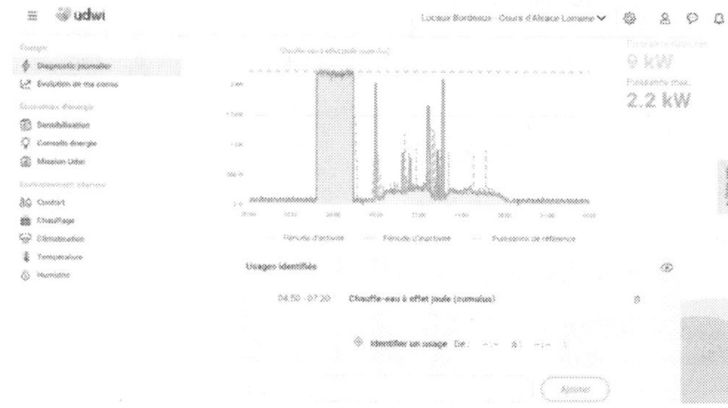

Figure 2. Screenshot of the interactive website showing the daily load curve and identified use periods.

A. Program deployment

The SEIZE program was built around 5 actions, each user being free to carry out one or more actions among:

- Environmental awareness workshops for managers and employees, aiming at raising attention about energy-saving actions in organizations;
- Installation of smart meters and comfort sensors, with access to the consumption data through an online visualization platform;
- Orientation towards financial aid schemes or technical advice and telephone/face-to-face appointment to redirect the professional to solutions adapted to their needs;
- Access to a network of ambassadors, who are professionals committed to promoting the program;
- Technical visits of buildings and delivery of recommendations concerning the management of the building or its energy renovation.

All these actions are offered to beneficiary companies and local authorities on a voluntary basis and free of charge. This paper concentrates on presenting the data collected by the smart meters installed during the SEIZE program. Datasets creation and analyses presented have been carried out by Eco CO2's Research and Development team and funded by Eco CO2.

B. Data visualization platform

To help organizations understand and reduce their electricity usage, users can visualize their consumption data in near real-time through an interactive website. They also receive weekly emails summarizing key trends in their consumption patterns. Figure 2 shows a screenshot of one of the website's pages displaying a daily load curve. Additional information, such as standby power over the last day, indoor and outdoor temperatures, indoor humidity, and the evolution of daily, weekly, and monthly electricity consumption, is also available, along with eco-gestures suggestions.

C. Sensors

Each participant of the study may have multiple sites, typically corresponding to individual buildings, each equipped with a unique electrical sensor. Most sites are also outfitted with one or more indoor comfort sensors measuring temperature and humidity. Outdoor temperature data are obtained using the building location [12].

Sensors' deployment took place between mid-2021 and end of 2024. At the time this paper is written, most sensors reported data for at least one year, as Figure 3 shows. Only sources with 4 weeks of data with sufficient quality have been kept in the database, which explains the flat lines from 0 to 4 weeks on Figure 3.

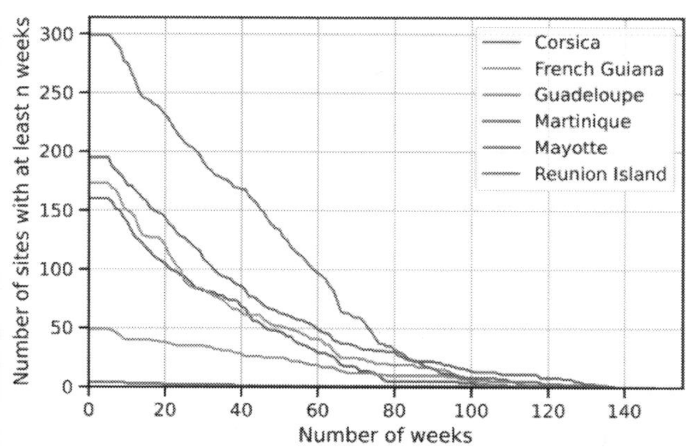

Figure 3. Evolution of the minimum number of weeks with electricity data in the dataset for each region.

D. Types of sensors

To facilitate sensor deployment, only sensors that do not require modifications of the users' electrical installations are used:

- Digital sensors communicate electronically with electricity meters to provide consumption indices.
- Analog sensors measure energy consumption by counting light pulses or wheel rotations from electromechanical meters.

Depending on the technology, the data timestep varies between one minute and one hour. External temperature data [12], indoor temperatures, and humidity levels are recorded every ten minutes.

III. DATASET

This section details the structure, processing, and anonymization of the dataset collected through the SEIZE program.

A. Data anonymization

To protect the users' privacy and follow the European General Data Protection Regulation (GDPR) guidelines [13], we resampled all the data to a weekly basis using `energy_analysis_toolbox`, a Python library developed by Eco CO2 R&D team and released in open-source [14]. Moreover, we provide for each participant its field of activity using the Statistical Classification of Economic Activities in the European Community (NACE) [15] classification.

B. Index data processing

The digital sensors transmit the consumption index, namely the integrated electricity consumption in kWh since the date of installation. Because of technical issues, abnormal index jumps are present in the database. Thus, we preprocess the data to automatically detect these anomalies. Weeks containing such elements are marked as invalid, and an empty entry replaces them in the `weekly_E_kWh` column of the dataset. This implies that the calculated energies can be considered valid on a weekly scale, and that the energy integrated over several weeks may be below reality if no data is provided for a week.

Furthermore, we interpolate between the last index of each week and the first known index of the following week to estimate weekly energy consumption as accurately as possible. If the time between the last known index and the first known index of the following week is too long, an empty entry is also written in the dataset.

C. Power data processing

The analog sensors detecting light pulses on electronic meters or wheel revolutions on electromechanical meters report the average power consumed each minute. The energy consumed is then calculated by taking the integral of the power. Weeks containing aberrant data, such as abnormally low or high power, abnormal load curve shapes, are removed from the analysis and an empty entry is written in the corresponding week of the dataset. Thus, similarly to digital sensors, calculated electricity consumption can be considered valid on a weekly scale, but energy integrated over several weeks may be below reality if no data is provided. In all following figures, we estimate the annual consumption by imputing missing week with the mean value consumed during the weeks kept in the dataset. Table II shows a few statistics about the dataset.

TABLE II. TOTAL NUMBER OF SITES NUMBER, CUMULATED WEEKS OF DATA, AND YEARLY ELECTRICITY CONSUMPTION STATISTICS, PER REGION. IQR STANDS FOR INTERQUARTILE RANGE.

Region	Sites	Weeks	Mean [MWh]	Min [MWh]	Max [MWh]	IQR [MWh]
Reunion Island	299	15362	46.9	0.5	517.9	48.4
Martinique	196	9729	38.1	0.6	383.3	42.4
French Guiana	173	7410	36.9	0.5	461.2	31.8
Corsica	160	6731	30.0	0.5	429.9	32.6
Guadeloupe	49	2860	38.5	0.6	436.8	50.7
Mayotte	4	196	12.2	3.1	22.0	5.3

D. Temperature and humidity data

A practical method for using outdoor temperature data alongside electricity consumption data is to calculate Heating Degree-Days (HDD) and Cooling Degree-Days (CDD). In the dataset, we compute these with several reference temperatures T_0 (15°C to 18°C for HDD and from 22°C to 26°C for CDD, both by steps of 1°C) using the integral method:

$$HDD_{T_0} = \int_{\text{Monday 0h}}^{\text{next Monday 0h}} [T_0 - T_{\text{ext}}(t)]^+ \, \mathrm{d}t, \quad (1)$$

$$CDD_{T_0} = \int_{\text{Monday 0h}}^{\text{next Monday 0h}} [T_{\text{ext}}(t) - T_0]^+ \, \mathrm{d}t, \quad (2)$$

where T_{ext} the external temperature, and:

$$x^+ = \begin{cases} x & \text{if } x > 0, \\ 0 & \text{otherwise.} \end{cases} \quad (3)$$

E. Dataset documentation

The dataset is available for download on data.gouv.fr [16] and Kaggle [17] as a single CSV file. In this file, each line corresponds to one week of data for a given site. Each line in the dataset represents data for a specific week and contains the following columns:

- `year-Wweek`: The ISO week. For example, "2023-W01" corresponds to the week from Monday, January 2 to Sunday, January 8, 2023.
- `user_id`: A unique identification number for each user.
- `site_id`: A unique identification number for each site. Each user can have multiple sites, but each site has precisely one electrical sensor. A site is typically a building, or a group of buildings.
- `department`: The name of the department where the sensor is located. Possibilities are `South Corsica`, `Upper Corsica`, `Guadeloupe`, `Martinique`, `Guyane`, `Reunion Island` and `Mayotte`.
- `nace_code`: The NACE code corresponding to the unique identifier of the company (SIRET) provided by the user (e.g., `47.71` for "Retail sale of clothing in specialized stores").
- `insee_code`: Unique numerical identifier assigned by the French National Institute of Statistics and Economic

Studies (INSEE) to French municipalities. On the January, 1st, January, 2023, France had 34945 municipalities [18].

- `weekly_E_kWh`: Weekly energy consumption in kWh.
- `weekly_dd_heating_<T0>`: Weekly HDD based on the reference temperature T_0, computed with Eq. 1.
- `weekly_dd_cooling_<T0>`: Weekly CDD based on the reference temperature T_0, computed with Eq. 2.
- `<stats_type>_indoor_<data_type>_<nn>`: Various statistics for indoor temperature and humidity data recorded by the comfort sensors. `<stats_type>` can be "min", "max" or "average". `<data_type>` can be "temperature" or "humidity". For temperature data, values are given in degrees Celsius, and for humidity, it is relative humidity in %. `<nn>` is a unique number for each comfort sensor between 00 and 99. The same number is used for temperature and humidity. For example, two columns named `min_indoor_temperature_00` and `min_indoor_humidity_00` correspond to the same physical sensor.

IV. Analysis and Visualization

This section presents the analysis and visualization of the dataset, focusing on electricity consumption patterns across different sectors and regions.

A. Estimated annual consumption by NACE category

To get insights on the annual consumption depending on the activity sector, we grouped companies in 9 categories based on their NACE code, as explained in Table III. As Figure 4 shows, there are significant consumption differences between the various sectors, with differences of median values for two sectors going up to an order of magnitude.

This discrepancy can be attributed to the nature of the activities in these sectors, where office work primarily involves low energy-consuming activities like lighting and computing, whereas shops and transport, as well as accommodation and catering, typically require more energy for operations, such as refrigeration, cooking, cooling, and heating.

TABLE III. Relationship between NACE sections and the categories used in Figure 4.

Category	NACE sections
Natural resource extraction and agriculture	A, B
Manufacturing and construction	C, D, E, F
Shops and transport	G, H
Accommodation and catering	I
Information and finance	J, K, L
Technical and administrative services	M, N
Public and social services	O, P, Q
Arts and recreation	R, S, T
Extra-territorial activities	U

Consumption varies significantly within each category, spanning more than two orders of magnitude between the minimum and maximum values. This may be due to other explanatory parameters not provided in the dataset, such as the number of employees, building surface area, number of equipment, equipment efficiency, or heating energy.

B. Thermosensitivity analysis

French islands and overseas regions studied in this paper exhibit diverse climatic conditions. Reunion Island, Mayotte, French Guiana, Martinique and Guadeloupe all experience tropical climates. However, while Reunion Island and Mayotte are both located in the southern hemisphere, Martinique and Guadeloupe are in the northern hemisphere, and French Guiana is very close to the Equator. Hence, the climate experienced in Reunion Island and Mayotte is relatively similar to that experienced in Martinique and Guadeloupe, but offset by 6 months because of the locations in different hemispheres.

Reunion Island and Mayotte are both characterized by hot and humid weather throughout the year, with distinct wet and dry seasons. The dry season takes place during May to

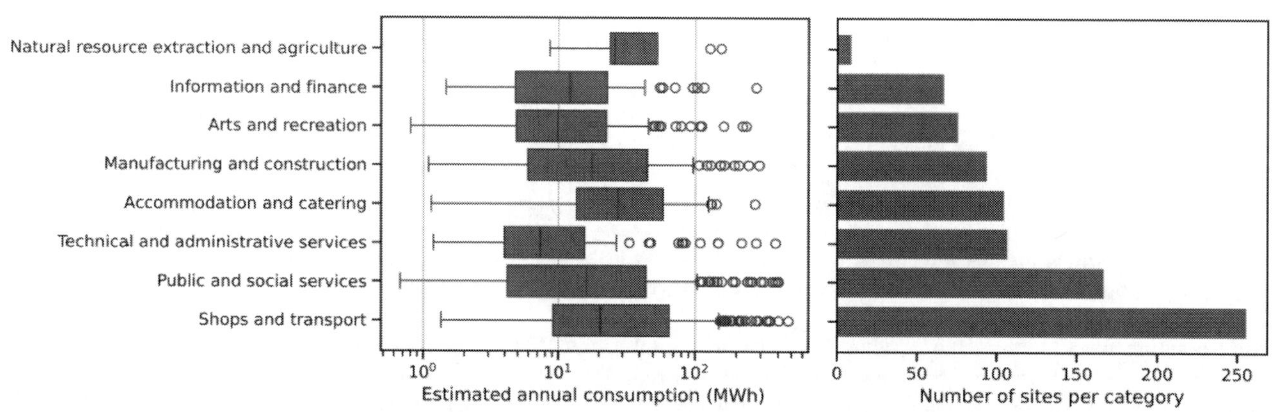

Figure 4. (left) Boxplot of the estimated annual consumption by company category. Whiskers are drawn at ±1.5 interquatile range (IQR), and all values outside this interval are represented with circles. (right) Number of sites per category.

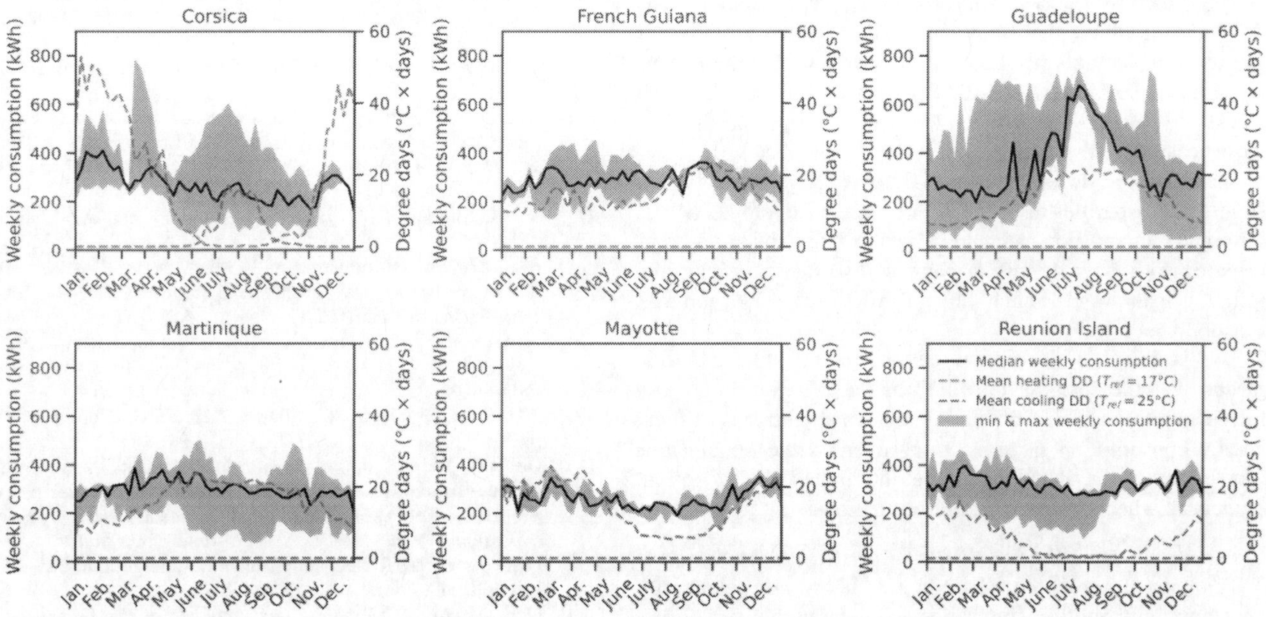

Figure 5. Seasonal evolution of the weekly consumption for each region. Each point represents one week. The black line represents the median weekly consumption across years and sites for each week of the year. For instance, the point for week 10 is the median consumption for all sites in a region across week 10 of 2021, 2022, 2023 and 2024. The gray-shaded area represents the minimum and maximum of the weekly consumption across sites, respectively. The blue and red dashed curves represent the mean HDD and CDD computed using Eq. 1 and Eq. 2 for heating (T_{ref} = 17°C) and cooling (T_{ref} = 25°C), respectively, averaged across years and sites.

November, when the weather is generally mild (typically 21 to 28°C) on the coast. The rainy season begins in December and ends in April. As Figure 5 shows, this translates into no need for heating throughout the year, and some cooling needs between November and April. This is reflected in the consumption, where a small correlation between mean CDD and usage can be seen.

Due to its proximity with the Equator, temperature and humidity in French Guiana varies very little throughout the year. Therefore, even if participants use air conditioning, the corresponding consumption remains roughly constant, which we indeed observe in Figure 5. It is also worth noting that consumption can vary greatly from one year to the other, depending on the territory, as the gray-shaded area in the same figure indicates.

Martinique and Guadeloupe share a similar tropical maritime climate. These islands experience warm temperatures year-round, with a wet season typically occurring from June to November, and a drier period from December to May. This can indeed be seen in the corresponding blue dashed curves in Figure 5. Interestingly, while the CDD curves are very similar for Martinique and Guadeloupe, we observe a much larger thermosensitivity in Guadeloupe than in Martinique. This suggests that, for the population studied, air conditioning systems are more used in Guadeloupe than in Martinique.

Corsica, in contrast, has a Mediterranean climate, characterized by hot, dry summers and mild, wet winters. This climate creates distinct heating and cooling needs, with a higher sensitivity to cooling during the summer compared to heating in the winter. This may be partially because air conditioning always requires electricity, while some heating systems do not. As a result, heating consumption may not be fully accounted for in Figure 5.

Finally, as Figure 6 shows, the distribution of consumption by region is very similar between all regions, the only exception being Mayotte, due to the fact that only 4 sensors are present in that region.

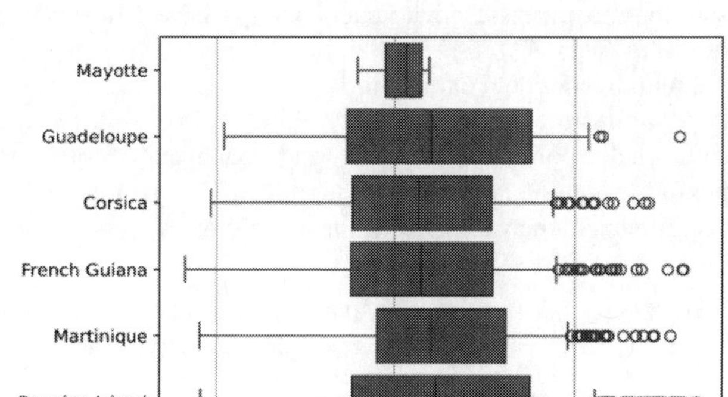

Figure 6. Boxplot of the estimated annual consumption by region. Whiskers are drawn at $\pm$1.5 IQR, and all values outside this interval are represented with circles.

C. Dataset limitations

The SECNIR dataset, while extensive and valuable, has certain limitations that must be acknowledged. Firstly, as explained in sections III-B and III-C, the dataset contains gaps for some weeks: over the 42288 weeks of the dataset, 5358 weeks (12.7%) do not contain electrical consumption data for these reasons.

Secondly, the dataset's territorial coverage is uneven, with significant disparities in the number of sensors deployed across different regions. For example, Mayotte is notably underrepresented with only 4 sensors compared to other regions like Reunion Island and French Guiana, which have 299 and 173 sensors respectively.

The scarcity of sensors in certain regions like Mayotte means that the insights drawn from these regions may not fully capture the variability and unique characteristics of local energy consumption patterns. Consequently, the conclusions derived from the data should be interpreted with caution, especially when comparing different regions.

V. Conclusion and Future Work

This paper presented SECNIR, a dataset on electricity consumption from 881 companies and local authorities across six French overseas regions. This dataset provides a valuable resource for analyzing consumption patterns in regions that are typically underrepresented in large-scale studies due to their isolated and tropical nature. By including electric, weather, and comfort data, this dataset allows for a nuanced understanding of how local climates influence electricity consumption.

The diverse climates of the regions studied—ranging from the tropical climates of Reunion Island, Mayotte, French Guiana, Martinique, and Guadeloupe to the Mediterranean climate of Corsica—highlight the importance of context-specific strategies in managing energy demand.

The findings from this study emphasize the need for tailored energy efficiency programs and policies that take into account the unique consumption patterns and climatic conditions of these regions. As these regions often rely heavily on fossil fuels, there is a pressing need to explore decarbonized energy solutions to mitigate environmental impacts and dependence on fossil fuels.

Future work could explore intra-day consumption variations, in a similar manner to the work done in [5]. This research could help policy-makers and grid managers identify the flexibility measures needed to support the transition to fully decarbonized energy mixes in these regions.

References

[1] M. Pullinger *et al.*, "The IDEAL household energy dataset, electricity, gas, contextual sensor data and survey data for 255 UK homes," en, *Scientific Data*, vol. 8, no. 1, p. 146, May 2021, ISSN: 2052-4463. DOI: 10.1038/s41597-021-00921-y.

[2] R. Medico *et al.*, "A voltage and current measurement dataset for plug load appliance identification in households," en, *Scientific Data*, vol. 7, no. 1, p. 49, Feb. 2020, ISSN: 2052-4463. DOI: 10.1038/s41597-020-0389-7.

[3] O. Parson *et al.*, "Dataport and NILMTK: A building data set designed for non-intrusive load monitoring," en, in *2015 IEEE Global Conference on Signal and Information Processing (GlobalSIP)*, Orlando, FL, USA: IEEE, Dec. 2015, pp. 210–214, ISBN: 978-1-4799-7591-4. DOI: 10.1109/GlobalSIP.2015.7418187.

[4] G. Liu *et al.*, "EWELD: A Large-Scale Industrial and Commercial Load Dataset in Extreme Weather Events," en, *Scientific Data*, vol. 10, no. 1, p. 615, Sep. 2023, ISSN: 2052-4463. DOI: 10.1038/s41597-023-02503-6.

[5] K. Bellinguer, R. Girard, A. Bocquet, and A. Chevalier, "ELMAS: A one-year dataset of hourly electrical load profiles from 424 French industrial and tertiary sectors," en, *Scientific Data*, vol. 10, no. 1, p. 686, Oct. 2023, ISSN: 2052-4463. DOI: 10.1038/s41597-023-02542-z.

[6] G. Gilboire *et al.*, "Energy balance Reunion Island 2021 + Key Figures 2021 (2022 edition)," France, Tech. Rep., 2022, INIS-FR–23-0083, p. 130.

[7] RTE France, "Electrical Balance 2023," Tech. Rep., Jul. 2024, p. 30.

[8] Court of Audit, "Support for non-interconnected zones," Court of Audit, Final observations S2023-1177, Apr. 2023, p. 5.

[9] French Energy Regulatory Commission, *Energy transition in the non-interconnected zones*, https://www.cre.fr/electricite/transition-energetique-dans-les-zni.html, , retrieved January, 2025.

[10] Court of Audit, *Complete cost analysis of the French power generation system*, Dec. 2021.

[11] Deliberation 2022-64 of the Energy Regulatory Commission on the draft decision on the incentive regulation framework for the Enedis advanced-metering system in the LV voltage field lower or equal to 36 kVA (Linky) for the period 2022-2024 and modifying deliberation no. 2021-13 related to the decision on the tariff for the use of public electricity distribution networks (TURPE 6 HTA-BT), Feb. 2022.

[12] Weather data provided by OpenWeather, https://openweathermap.org, retrieved January, 2025.

[13] Official Journal of the European Union, Regulation (EU) 2016/679 of the European Parliament and of the Council of 27 April 2016 on the protection of natural persons with regard to the processing of personal data and on the free movement of such data, and repealing Directive 95/46/EC (General Data Protection Regulation)., 2016.

[14] EcoCO2, *Energy_analysis_toolbox*, https://github.com/Eco-CO2/energy_analysis_toolbox, retrieved January, 2025., 2024.

[15] EUROSTAT, "NACE Rev. 2," Tech. Rep., 2008.

[16] P.-A. Chevreuil, A. Tavant, P. Costini, C. Simi, and J. Costes, *SECNIR: Electricity consumption of businesses and local authorities in non-interconnected French regions*, https://www.data.gouv.fr/fr/datasets/consommation-delectricite-des-entreprises-et-collectivites-locales-dans-les-regions-francaises-non-interconnectees-1/, retrieved January, 2025., Sep. 2024.

[17] P.-A. Chevreuil, A. Tavant, P. Costini, C. Simi, and J. Costes, *SECNIR: Electricity consumption of businesses and local authorities in non-interconnected French regions*, https://www.kaggle.com/datasets/ecoco2/secnir-consumption-of-french-overseas-companies, retrieved January, 2025., Sep. 2024.

[18] L. Sebbane, *Statistical information bulletin 172 of the General Directorate of Local Authorities*, Mar. 2023.

Disaggregation of Heating and Cooling Energy Consumption via Maximum a Posteriori Estimation

Antoine Tavant
Cédric Simi
Eco CO2
Lyon, France
e-mail: antoine.tavant@m4x.org

Pierre-Alexis Chevreuil
Pierre Costini
Eco CO2 Nice, France
e-mail: pierre.costini@centraliens.net
e-mail: pierrealexis.chevreuil@gmail.com

Joris Costes
Eco CO2 and *Univ. Paris-Saclay, ENS Paris-Saclay, CNRS, Centre Borelli,*
Gif-sur-Yvette, France
e-mail: joris.costes@ens-paris-saclay.fr

***Abstract*—Estimating energy use in heating and air-conditioning systems is crucial for effective building energy management. This article introduces a new method combining the use of degree-days with the maximum a posteriori estimation statistical method to disaggregate heating and cooling energy consumption from other uses. Degree-days provide a reliable measure of the demand for energy needed to heat or cool a building, while a posteriori estimation offers a robust statistical approach to refine these estimates based on available data. A significant challenge addressed by this method is the need to accurately estimate the parameters of the model, which is achieved here by leveraging a comprehensive database. The method's efficacy is demonstrated through a case study of a building with one year of collected data, illustrating its practical application. Our findings underscore the method's potential to enhance energy management practices and guide future research in heating and cooling energy estimation.**

Keywords-Smart meters; Non-Intrusive Load Monitoring; NILM; Thermosensitivity.

I. INTRODUCTION

With the increasing deployment of smart meters in buildings, it is now possible to access aggregated energy usage data with unprecedented ease and accuracy [1]. This wealth of data opens up new opportunities for advanced analytical methods to improve energy estimation practices.

Smart meters usually collect the total building energy consumption data at a very low temporal resolution. For instance, in France the daily consumption is the only quantity provided by default by communicating meters, with finer sampling rate mainly depending on the supply contract and meter type. Individual users have the possibility to opt in to a finer 30 minutes sampling rate, but it is rarely selected. These low frequencies increase the difficulty to detect appliances [2]. Usual disaggregation techniques discard explanatory variables, such as weather and date, to focus on pattern and signature detection [2].

The management of energy use in Heating, Ventilation, and Air Conditioning (HVAC) systems is a critical aspect of building operations, influencing both cost efficiency and environmental impact. HVAC represent 38% of buildings consumption worldwide, and up to 60% in Europe [1]. Hence, accurate estimation of heating and cooling energy consumption is essential for optimizing energy use, reducing costs, and meeting sustainability goals. There are two families of methods to model building energy consumption [3]:

- **The forward methods** use the building description to develop a theoretical energy signature model,
- **The reverse methods** use collected consumption data to fit a consumption model.

Forward models are usually used when few or no data is available, such as during design and during the first few months of use. However, it has been shown that they show large discrepancies when compared to actual consumption, as the occupants' behavior can have a significant impact on the consumption [4]. This article focuses on reverse methods, available when a significant amount of data has been gathered.

In France, 37% of households use electricity for heating [5]. On top of that, the air conditioning energy use is expected to increase in the future [6]. Hence, the assessment of HVAC energy uses with smart-meters requires disaggregating both the cooling and the heating energy uses from the total consumption measured.

Figure 1 introduces the typical data used in this study. The data have been collected in one household near Lyon, France, with electrical heating and no cooling system. The energy consumption is gathered using the Linky smart-meter of Enedis [7]. The OpenWeatherMap API was used for the outdoor temperature [8]. The data is aggregated on a weekly frequency, from Monday to Sunday, to reduce the impact of the day of the week on energy consumption, considering the differences between workdays and weekends. The top panel of Figure 1 shows the temporal evolution of the total weekly electric energy consumed by the household. The bottom panel of Figure 1 presents the same data as the left panel, showing the outside temperature to energy consumption relation. One point in black corresponds to the week starting on Monday the 1st of January 2024, a week when occupants were not present. We can identify two regions:

- when the mean weekly outside temperature is above 16°C, the energy consumption seems independent of the temperature,
- when the mean weekly outside temperature is below 16°C, the energy consumption seems negatively proportional to the temperate.

This article introduces an innovative method for estimating HVAC energy use that leverages Degree-Days (DD) computation combined with Maximum a Posteriori (MAP) estimation

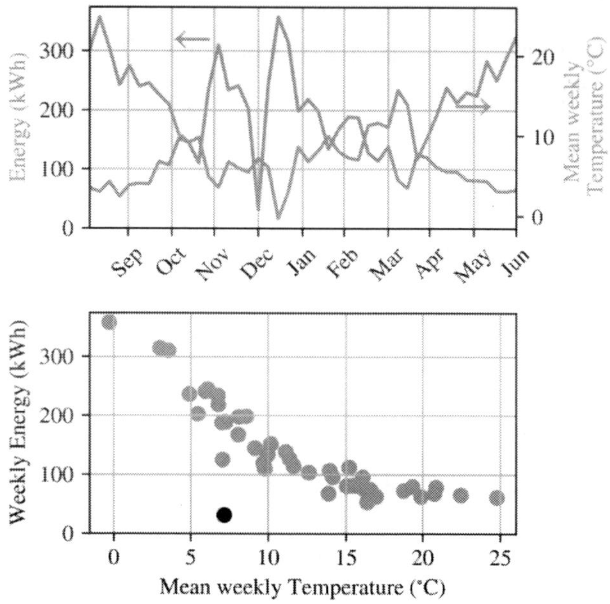

Figure 1. Data collected for one household: (top) the temporal evolution of the weekly energy (left axis) and the mean weekly outside temperature (right axis). (bottom) the weekly energy as a function of the mean weekly temperature. One point in black corresponds to an outlier.

which is more accurate than simpler approach such as a threshold when electricity is used for heating as well as other uses. DD is a well-established metric that quantifies the demand for energy to heat or cool a building, based on temperature deviations from a baseline [4]. MAP estimation, on the other hand, is a statistical technique that refines these estimates by incorporating prior knowledge and observed data to produce the most probable outcomes. By combining these two tools, we provide a way to estimate the energy usage of HVAC once enough data is collected.

To introduce the concept behind MAP estimation, consider the following analogy. Suppose a player rolls two dice and the goal is to determine the individual values of each dice. Without prior knowledge, this task is impossible. However, knowing that each dice can show a number between one and six, with a uniform probability distribution, the total number of possible outcomes is $\frac{6\times7}{2} = 21$, giving a 4.76% chance of guessing the correct pair. If the player provides the combined sum of the two dice, the estimation becomes significantly more manageable. For sums of 2, 3, 11, or 12, the possible combinations of the dice values are unique and known with certainty. Conversely, for sums of 6, 7, or 8, there are three possible pairs, resulting in a $\frac{1}{3}$ probability of guessing the correct pair. This analogy illustrates that by integrating prior knowledge (the uniform probability distributions of each dice) with an observation (the sum of the dice), the a posteriori estimation of each dice's value is substantially improved compared to the a priori estimation. In this study, the observation is the aggregated energy consumption recorded by a smart meter, which represents the sum of heating, cooling, and other energy uses. By applying MAP estimation, we can more accurately decompose this sum into its constituents, thereby enhancing the precision of our energy usage estimates.

A notable challenge in implementing this method is the need to estimate the prior distributions of the heating, cooling, and other uses needs, and whether they are correlated or not. This correlation is essential for precise estimation but can be difficult to determine directly. To overcome this limitation, we utilize a comprehensive database that provides the necessary correlation information, ensuring that the estimation method remains robust and reliable.

The following sections of this article detail the components and implementation of this new estimation method. We begin in Section II with an overview of DD computation and MAP estimation, followed by a discussion on addressing the correlation of random variables. Section III outlines the step-by-step process of integrating these components. Section IV aims to estimate a correlation coefficient from an extensive dataset. We then present in Section V a case study applying the method to the data shown in Figure 1, demonstrating its practical application and effectiveness. Finally, we discuss in Section VI the results, compare them with traditional methods, and offer insights into the implications and potential future developments.

II. Thermosensitivity Consumption Model

In this section, we introduce the thermosensitivity consumption model, which forms the foundation of our method for estimating HVAC energy use. The total energy consumed in a building over a given period, E, can be decomposed into three primary components:

$$E = E_{\text{h}} + E_{\text{c}} + E_{\text{o}}, \tag{1}$$

where E_{h} is the energy required for heating, E_{c} is the energy required for cooling, and E_{o} represents all other energy uses not dependent on temperature variations, such as lighting, appliances, and other equipment.

A. Degree-Days

DD is a widely used metric to quantify the demand for energy needed to heat or cool a building [9]. Heating Degree-Days (HDD) and Cooling Degree-Days (CDD) are calculated based on deviations from a baseline temperature. The baseline temperatures depend on the building and occupants, and they need to be calibrated to maximize the performance of the model using historical data [9]. The duration of the integration period is at least one day, but can last longer. Practically, using a weekly sampling for the energy and the degree-days allows merging the differences related to the day of the week, such as weekends. HDD is used to estimate the energy needed to heat buildings. The integral formula to compute HDD is:

$$HDD = \int [T_{\text{base,h}} - T(t)]^{+} \, \text{d}t, \tag{2}$$

where $T_{\text{base,h}}$ is the heating baseline temperature, $T(t)$ is the outdoor temperature at time t, and $[\cdot]^+$ is the positive function

$$x^+ = \begin{cases} x & \text{if } x > 0, \\ 0 & \text{if } x \leq 0. \end{cases} \tag{3}$$

Equation (2) assumes that the heating is switched on when the outside temperature is below $T_{\text{base,h}}$. CDD are the equivalent of HDD for cooling needs. Its integral formulation is:

$$CDD = \int [T(t) - T_{\text{base,c}}]^+ \, \mathrm{d}t, \tag{4}$$

where $T_{\text{base,c}}$ is the cooling baseline temperature. It's worth noting that other ways to compute HDD and CDD have been proposed in the literature. While we chose to use the integral formulation, other formulations may be used without changing the rest of the methodology or conclusions of the paper.

B. Degree-Days and Energy Consumption

Using the notation H for the Heaviside step function defined as

$$H(x) = \begin{cases} 1 & \text{if } x > 0, \\ 0 & \text{if } x \leq 0, \end{cases} \tag{5}$$

the energy required for heating and cooling can be modeled linearly with respect to HDD and CDD, respectively:

$$E_{\text{h}} = (\alpha_{\text{h}} \cdot HDD + \epsilon_{\text{h}})\, H(HDD), \tag{6}$$

$$E_{\text{c}} = (\alpha_{\text{c}} \cdot CDD + \epsilon_{\text{c}})\, H(CDD), \tag{7}$$

while other uses' energy consumption is modeled by

$$E_{\text{o}} = E_{\text{baseline}} + \epsilon_{\text{o}}, \tag{8}$$

where α_{h} and α_{c} are coefficients that represent the sensitivity of energy use to DD, and ϵ_{h}, ϵ_{c} and ϵ_{o} are random variables that capture the deviation from the model for heating, cooling, and other uses, respectively. The Heaviside step function ensures that ϵ_{h} and ϵ_{c} are added when the corresponding use is required.

III. Methodology for HVAC consumption disaggregation

This section describes the new proposed method to disaggregate the cooling and heating energy from the other uses.

A. Consumption model

The deviations ϵ_{h}, ϵ_{c} introduced in (7) and (6) account for various uncertainties and factors that affect energy consumption, but are not linearly correlated to the DD. These can include variations in building occupancy, operational schedules, sun illumination, and efficiency of the HVAC systems. Additionally, ϵ_{o} of (8) represents the variation in the energy use for non-HVAC purposes. Notably, they do not represent measurement error, which we assume to be nonexistent.

It is important to note that these variables can be correlated. For instance, an unusual higher occupancy of the building may increase the energy consumed by HVAC, as well as the energy consumed by other uses. On the other hand, a change in the AC temperature settings will increase the energy use for cooling, but not the one concerning other uses. We assume the deviations to the thermosensitivity model with normal distributions:

$$[\epsilon_{\text{h}}, \epsilon_{\text{c}}, \epsilon_{\text{o}}]^T = \mathcal{N}(\mathbf{0}, \mathbf{\Sigma}), \tag{9}$$

with $\mathbf{\Sigma}$ the covariance matrix. We assume in the following that the sampling period is short enough so that heating and cooling cannot be present within the selected period:

$$\begin{cases} HDD = 0 & \text{if } CDD > 0, \\ CDD = 0 & \text{if } HDD > 0. \end{cases} \tag{10}$$

This hypothesis typically requires the sampling period to be shorter than a month. Hence, we can express the covariance matrix as

$$\mathbf{\Sigma} = \begin{pmatrix} \sigma_{\text{h}}^2 & 0 & \rho_{\text{h}}\sigma_{\text{h}}\sigma_{\text{o}} \\ 0 & \sigma_{\text{c}}^2 & \rho_{\text{c}}\sigma_{\text{c}}\sigma_{\text{o}} \\ \rho_{\text{h}}\sigma_{\text{h}}\sigma_{\text{o}} & \rho_{\text{c}}\sigma_{\text{c}}\sigma_{\text{o}} & \sigma_{\text{o}}^2 \end{pmatrix}, \tag{11}$$

where σ_x is the standard deviations of ϵ_x, with $x \in \{h, c, o\}$, and ρ_{h} (respectively ρ_{c}) represents the correlation coefficient between the ϵ_{h} (respectively ϵ_{c}) and ϵ_{o}. In the following section, we will detail the probabilistic inference method proposed to estimate the values of ϵ_{h}, ϵ_{c} and ϵ_{o}.

B. Inference

The goal of this section is to outline the method for estimating E_{h}, E_{c}, and E_{o} from a single measurement of total energy consumption E, given the HDD, CDD, and the prior distributions (9). Combining (1) with (7) to (8) and (10), we have:

$$E = E_{\text{baseline}} + \epsilon_{\text{o}} + \begin{cases} \alpha_{\text{c}} \cdot CDD + \epsilon_{\text{c}} & \text{if } CDD > 0, \\ \alpha_{\text{h}} \cdot HDD + \epsilon_{\text{h}} & \text{if } HDD > 0. \end{cases} \tag{12}$$

Determining the distribution of energy consumption between heating, cooling and other uses means that we have to estimate the values of $\epsilon_{\text{o}}, \epsilon_{\text{c}}$ and ϵ_{h}. As we assumed previously that the sampling period was chosen short-enough to ensure that heating needs are null when there are cooling needs and vice versa, we can study either of the 2 cases, and the result will be translatable to the other case. Thus, we start with the first case $HDD > 0$. The exact same derivation can be followed for the other case $CDD > 0$.

The total deviation of the measured energy E from the thermosensitivity model is

$$res = E - E_{\text{baseline}} - \alpha_{\text{h}} \cdot HDD = \epsilon_{\text{o}} + \epsilon_{\text{h}} \tag{13}$$

Hence, res is a random variable following the distribution

$$res \sim \mathcal{N}(0, \sigma_{\text{h+o}}), \tag{14}$$

with

$$\sigma_{\text{h+o}}^2 = \sigma_{\text{o}}^2 + \sigma_{\text{h}}^2 + 2\rho_{\text{h}}\sigma_{\text{h}}\sigma_{\text{o}}. \tag{15}$$

The MAP estimation of $\epsilon_{\text{h}}, \epsilon_{\text{o}}$ is defined as

$$\hat{\epsilon}_{\text{h}}, \hat{\epsilon}_{\text{o}} = \underset{\epsilon_{\text{h}}, \epsilon_{\text{o}}}{\arg\max}\, P(\epsilon_{\text{h}}, \epsilon_{\text{o}} | res), \tag{16}$$

with $\hat{x}$ the estimation of x and P the probability density function. Equation (16) means that the MAP estimation of ϵ_h, ϵ_o are the values that maximize the likelihood function of (ϵ_h, ϵ_o), knowing res. Using the Bayes theorem,

$$P(\epsilon_h, \epsilon_o|res) = \frac{P(res|\epsilon_h, \epsilon_o)P(\epsilon_h, \epsilon_o)}{P(res)}. \tag{17}$$

Hence, as $P(res)$ doesn't depend on ϵ_h, ϵ_o:

$$\hat{\epsilon}_h, \hat{\epsilon}_o = \arg\max_{\epsilon_h, \epsilon_o} P(res|\epsilon_h, \epsilon_o)P(\epsilon_h, \epsilon_o). \tag{18}$$

Recalling (13), (18) becomes:

$$\hat{\epsilon}_h, \hat{\epsilon}_o = \arg\max_{\epsilon_h, \epsilon_o} P(\epsilon_h, \epsilon_o). \tag{19}$$

The join probability density function of the bi-variable distribution

$$\begin{bmatrix}\epsilon_h \\ \epsilon_o\end{bmatrix} \sim \mathcal{N}\left(\begin{bmatrix}0 \\ 0\end{bmatrix}, \begin{bmatrix}\sigma_h^2 & \rho_h\sigma_h\sigma_o \\ \rho_h\sigma_h\sigma_o & \sigma_o^2\end{bmatrix}\right) \tag{20}$$

is of the form

$$P(\epsilon_h, \epsilon_o) \propto \exp\left(-\frac{1}{2(1-\rho_h^2)}\left(\frac{\epsilon_h^2}{\sigma_h^2} + \frac{\epsilon_o^2}{\sigma_o^2} - \frac{2\rho_h\epsilon_h\epsilon_o}{\sigma_h\sigma_o}\right)\right). \tag{21}$$

Substituting (21) and (13) in (19), we obtain

$$\hat{\epsilon}_h = \arg\min_{\epsilon_h}\left(\frac{\epsilon_h^2}{\sigma_h^2} + \frac{(res-\epsilon_h)^2}{\sigma_o^2} - \frac{2\rho_h\epsilon_h(res-\epsilon_h)}{\sigma_h\sigma_o}\right), \tag{22}$$

$$\hat{\epsilon}_o = \arg\min_{\epsilon_o}\left(\frac{(res-\epsilon_o)^2}{\sigma_h^2} + \frac{\epsilon_o^2}{\sigma_o^2} - \frac{2\rho_h(res-\epsilon_o)\epsilon_o}{\sigma_h\sigma_o}\right). \tag{23}$$

That can be solved using the quadratic formula, to

$$\hat{\epsilon}_x = res\frac{\sigma_x^2 + \rho_h\sigma_o\sigma_h}{\sigma_o^2 + \sigma_h^2 + 2\rho_h\sigma_o\sigma_h}, \tag{24}$$

with x either h or o.

Thus, the MAP estimation of the heating energy $\hat{E}_h$ and of the other uses $\hat{E}_o$ is

$$\hat{E}_o = E_{baseline} + res\frac{\sigma_o^2 + \rho_h\sigma_o\sigma_h}{\sigma_o^2 + \sigma_h^2 + 2\rho_h\sigma_o\sigma_h}, \tag{25}$$

$$\hat{E}_h = \alpha_h \cdot HDD + res\frac{\sigma_h^2 + \rho_h\sigma_o\sigma_h}{\sigma_o^2 + \sigma_h^2 + 2\rho_h\sigma_o\sigma_h}. \tag{26}$$

A way to understand this estimation is to say that the deviation from the linear model, noted res, is split in two parts, one attributed to the heating energy usage, and the second to the other uses. The relative importance of the two parts depends on the relative value of the standard deviations σ_h and σ_o and the correlation coefficient ρ_h. The same process is followed to derive the estimation during a period with $CDD > 0$.

C. *Estimation of the model parameters*

In order to apply the inference proposed previously, the values of the different parameters are needed. They depend on the building details, HVAC system and the behavior of occupants. Hence, they need to be estimated from measured data following a reverse model approach [4].

a) Other uses: We can estimate $E_{baseline}$ and σ_o from historical data when both $HDD = 0$ and $CDD = 0$:

$$\hat{E}_{baseline} = \langle E\rangle_{HDD=0,CDD=0} = \frac{1}{N}\sum_{i=1}^{N} E_i, \tag{27}$$

$$\hat{\sigma}_o^2 = \frac{1}{N-1}\sum_{i=1}^{N}(E_i - \hat{E}_{baseline})^2. \tag{28}$$

with N the number of data points where $HDD = 0$ and $CDD = 0$.

b) Thermosensitivity models: We can estimate σ_h and α_h from the periods where $HDD > 0$. The estimation of α_c and σ_c follows the same logic with the periods where $CDD > 0$. A linear model is fitted such that

$$E - E_{baseline} \sim \alpha_h \cdot HDD, \tag{29}$$

to find $\hat{\alpha_h}$ the estimation of α_h. The standard deviation of the residuals of the linear model res is estimated with

$$\hat{\sigma}_{h+o}^2 = \frac{1}{N_h - 1}\sum_{i=0}^{N_h}(E_i - E_{baseline} - \alpha_h \cdot HDD_i)^2, \tag{30}$$

with N_h the number of points with $HDD > 0$. We can then use (15) to find σ_h.

However, the value of ρ_h is unobservable, as neither E_h nor E_o are measured. Two options are available:

- estimate σ_h with the extreme values $\rho_h = -1$, $\rho_h = 0$ and $\rho_h = 1$ to compute a lower and upper bounds,
- use a benchmark value estimated elsewhere, on a similar building, for instance.

The next section investigates the second possibility to obtain an insight on ρ_h.

IV. Estimation of Residuals Correlation

In this section, we use the extensive database EDRP [10] to estimate the value of ρ_h. This database includes more than 8000 households in the United Kingdoms with both electric and gas energy consumption measured hourly between 2007 and 2011. From this dataset, 2319 households have been selected on the following criteria

- electricity consumption independent with HDD (linear regression with $p - value > 0.1$);
- gas consumption showing no significant outliers;
- more than 52 weeks of data available.

The first criteria mean that we can assume that the gas energy is mostly consumed for heating: E_h, and the electric energy consumption is related to the other uses: E_o. The two other criteria are used to improve the quality of the analysis. Figure 2 illustrates the analysis of one household. Figure 2(**a**) shows the available data, with the gas and electricity weekly energy, and HDD. Figure 2(**b**) shows the thermosensitivity models, and Figure 2(**c**) the relation between the gas and electricity residuals. From the residuals obtained, and assuming the gas is only used for heating and the electricity is only used for cooling, this household shows a correlation $\rho_h = 0.27$. Figure 3 shows the distribution of the correlations between

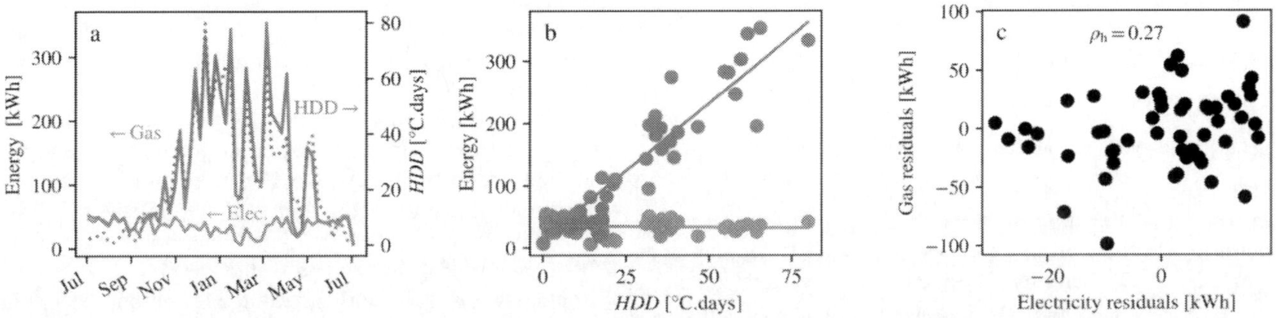

Figure 2. Illustration of one of the 2319 households analyzed. **a** : Time series of the gas (green) and electricity (red) weekly energy consumption (left axis), and HDD (blue dotted line, right axis). **b** : gas (green) and electricity (red) weekly energy consumption as function of HDD, markers represent the measured data showed in subplot **a**, and the solid lines represent the thermosensitivity models. **c**: scatter plot of the gas and electricity residuals to the models, meaning the difference between the markers and the lines of the subplot **b**. We compute for this household $\rho_{\mathrm{h}} = 0.27$.

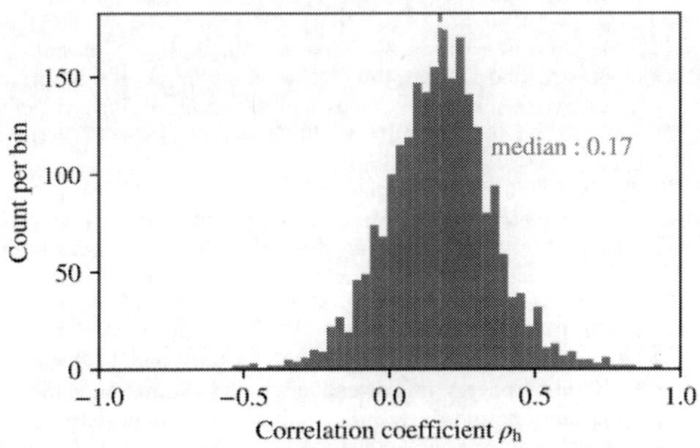

Figure 3. Distributions of the correlation between the electric and gas energy consumption residuals of the models, as illustrated in Figure 2, to estimate ρ_{h}. A total of 2319 households has been analyzed.

the residuals for a selection of the 2319 selected households. As we can see, the values are spread over a large domain, with values of ρ_{h} ranging from -0.54 to 0.93. Fortunately, the distribution is relatively narrow, with most households exhibiting a weak positive correlation between ϵ_{h} and ϵ_{o}. This means that using the median value $\rho_{\mathrm{h}} = 0.17$ is a relatively correct assumption for most cases.

V. Applications on a real building

In this section, we present the application of the proposed estimation to provide the disaggregation of the energy consumption for the data introduced in Figure 1. The HDD are computed with a reference temperature $T_{base,heating} = 17$°C. From the data, and using the median correlation coefficient measured in Section IV $\rho_{\mathrm{h}} = 0.17$, we can estimate that over one year:

- $E_{\mathrm{baseline}} = 69$ kWh
- $\sigma_{\mathrm{o}} = 8.9$ kWh
- $\alpha_{\mathrm{h}} = 1.8$ kWh/°C.week
- $\sigma_{\mathrm{h+o}} = 75$ kWh hence $\sigma_{\mathrm{h}} = 73$ kWh

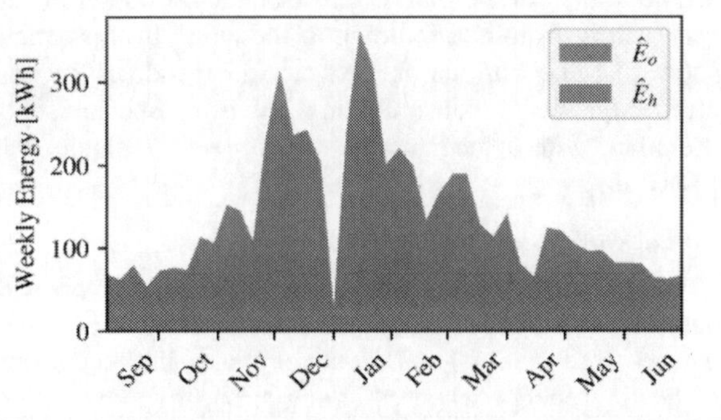

Figure 4. Disaggregation of the energy consumption of the household, presented in Figure 1: The orange area represents the Other uses, while the green area represent the heating energy. The value of $\rho_{\mathrm{h}} = 0.17$ is used.

Figure 4 presents the MAP estimation for the household whose data are shown in Figure 1 using the parameters estimated previously. We can see that $\hat{E}_{\mathrm{o}}$ remains relatively constant over the year, while $\hat{E}_{\mathrm{h}}$ increases in winter.

VI. Discussion

Several aspects of the method are discussed here-after.

A. Gains compared to alternative approaches

The usual approach of using a thermosensitivity model to estimate the heating and cooling energy needs is to discard the deviations to the model following the Maximum Likelihood Estimation (MLE) [11] :

$$\hat{E}_{\mathrm{o}MLE} = E_{\mathrm{baseline}} \tag{31}$$

$$\hat{E}_{\mathrm{h}MLE} = \alpha_{\mathrm{h}} \cdot HDD \tag{32}$$

$$\hat{E}_{\mathrm{c}MLE} = \alpha_{\mathrm{c}} \cdot CDD \tag{33}$$

This is an a priori estimation: the measurement of E is not used to update the estimates of the different components.

This approach has a Standard Error (SE) of $\sigma_{\mathrm{o}}, \sigma_{\mathrm{h}}$, and σ_{c} for $\hat{E}_{\mathrm{o}MLE}, \hat{E}_{\mathrm{h}MLE}$ and $\hat{E}_{\mathrm{c}MLE}$, respectively. One significant drawback is that $\hat{E}_{\mathrm{o}MLE} + \hat{E}_{\mathrm{h}MLE} + \hat{E}_{\mathrm{c}MLE} \neq E$, meaning that the disaggregation does not conserve energy.

On the other hand, we can show that the SE of the MAP estimations are

$$SE^2_{\hat{E}_x} = \frac{\left(\sigma_x^2 + \rho_x\sigma_o\sigma_x\right)^2}{\sigma_o^2 + \sigma_x^2 + 2\rho_x\sigma_o\sigma_x} < \sigma_o^2 \quad (34)$$

$$SE^2_{\hat{E}_o} = \frac{\left(\sigma_o^2 + \rho_x\sigma_o\sigma_x\right)^2}{\sigma_o^2 + \sigma_x^2 + 2\rho_x\sigma_o\sigma_x} < \sigma_x^2 \quad (35)$$

with x being h or c. In general, we have σ_o smaller than σ_x, meaning that the standard error of $\hat{E}_c$ and $\hat{E}_h$ are improved compared to the a priori estimates (32) and (33). On the other hand, $SE_{\hat{E}_o}$ is bound by σ_x which is usually larger than σ_o, hence $\hat{E}_o$ can potentially be worsened. This is due to the fact that the proposed a posteriori method validates (1), while the a priori estimation (31) does not, leading to the violation of the energy conservation.

B. Limitation due to the correlation coefficients

The values of ρ_h and ρ_c are difficult to obtain. They can vary wildly from one building to the other. In this article, we show that the correlation between the residuals of gas and electricity energy consumption vary from one household to the other, with a median value of $\rho_h = 0.17$ obtained in the EDRP dataset.

C. Lack of validation

To the best of our knowledge, there are no large open source datasets with labeled energy uses for heating or cooling. The dataset used in Section IV only provides indirect estimation of heating energy when assuming strong hypotheses, and no cooling energy is available. Hence, it was not possible to measure the performance of the proposed method.

D. Impact of occupancy on the consumption

The proposed method considered the outside temperature as the only explanatory variable. However, the occupancy and other factors such as price of electricity can also impact the building consumption. Such factors could be included in a more elaborate model, using direct values or proxies of these factors. For example, occupancy variation could be approximated using the day of the week, since most households have a different occupancy behavior during weekdays and weekends. It is also possible to estimate the occupancy from the energy consumption data itself [11].

VII. Conclusion and Future Work

In this article, we introduced a novel method for estimating the energy use of HVAC systems in buildings, leveraging degree-days computation and Maximum a Posteriori (MAP) estimation. By decomposing the total energy use into heating, cooling, and other components, and incorporating the impact of uncertainties through correlation coefficients, we provide a method to estimate the energy consumed by each component.

We outlined a systematic approach to implement this method, starting with the estimation of non-HVAC energy use from historical data points where there is no heating or cooling demand. We then demonstrated how to estimate the thermosensitivity coefficients and standard deviations of the heating and cooling energy components. Finally, we utilized correlation coefficients to derive MAP estimates for heating, cooling, and other energy uses, ensuring that the total energy consumption aligns with observed data.

The practical application of this method was illustrated through a case study, showcasing its potential for enhancing building energy management. The case study, based on the measurement of the heating correlation coefficient ρ_h from a dataset of more than two thousand households located in the UK, demonstrated the method's efficacy.

Future work will focus on validating the approach with a labeled dataset that includes both heating and cooling energy. Additionally, measuring the cooling correlation coefficient ρ_c and incorporating other explanatory variables, such as the day of the week to account for occupancy effects, are expected to further improve model performance.

References

[1] M. González-Torres, L. Pérez-Lombard, J. F. Coronel, I. R. Maestre, and D. Yan, "A review on buildings energy information: Trends, end-uses, fuels and drivers," en, *Energy Reports*, vol. 8, pp. 626–637, Nov. 2022, ISSN: 23524847. DOI: 10.1016/j.egyr.2021.11.280.

[2] A. Petralia, P. Charpentier, P. Boniol, and T. Palpanas, "Appliance Detection Using Very Low-Frequency Smart Meter Time Series," en, in *Proceedings of the 14th ACM International Conference on Future Energy Systems*, arXiv:2305.10352 [cs, eess], Jun. 2023, pp. 214–225. DOI: 10.1145/3575813.3595198.

[3] M. Mehregan, A. Naminezhad, S. Vakili, and M. Delpisheh, "Building energy model validation and estimation using heating and cooling degree days based on accurate base temperature," en, *Energy Science & Engineering*, vol. 10, no. 9, pp. 3638–3649, 2022, ISSN: 2050-0505. DOI: 10.1002/ese3.1246.

[4] G. Matheron, "Large-scale validation of energy performance diagnostics via consumption measured by connected meters," fr, *Hello Watt*,

[5] *Key figures of housing*. The Data and Statistical Studies Service (SDES), 2022.

[6] D. F. Birol, "The Future of Cooling," en, *IEA*, 2018.

[7] Deliberation 2022-64 of the Energy Regulatory Commission on the draft decision on the incentive regulation framework for the Enedis advanced-metering system in the LV voltage field lower or equal to 36 kVA (Linky) for the period 2022-2024, Feb. 2022.

[8] OpenWeatherMap, https://openweathermap.org, retrieved: January 2025.

[9] J. A. Azevedo, L. Chapman, and C. L. Muller, "Critique and suggested modifications of the degree days methodology to enable long-term electricity consumption assessments: A case study in Birmingham, UK," en, *Meteorological Applications*, vol. 22, no. 4, pp. 789–796, 2015, ISSN: 1469-8080. DOI: 10.1002/met.1525.

[10] AECOM Building Engineering, *Energy Demand Research Project: Early Smart Meter Trials, 2007-2010. [data collection]. UK Data Service. SN: 7591*, 2014. DOI: http://doi.org/10.5255/UKDA-SN-7591-1.

[11] F. Culière, L. Leduc, and A. Belikov, "Bayesian model of electrical heating disaggregation," in *Proceedings of the 5th International Workshop on NILM*, Nov. 2020, pp. 25–29. DOI: 10.1145/3427771.3427848.

Analysis of Data from a 5G-Based Photovoltaic Plant Monitoring System

Sofia D. Carvalho
Telematics Department
Federal Institute of Technology of Ceará (IFCE)
Fortaleza, Ceará, Brazil
Email: carvalhosofiadiniz@outlook.com

Paulo C. M. Carvalho
Department of Electrical Engineering
Federal University of Ceará (UFC)
Fortaleza, Brazil
Email: carvalho@dee.ufc.br

Luis M. Fernández-Ramírez
Research Group in Sustainable and Renewable Electrical Technologies (SURET), Department of Electrical Engineering, University of Cadiz (UCA), Algeciras, Spain.
Email: luis.fernandez@uca.es

Paulo R. X. da Silva
Telematics Department
Federal Institute of Technology of Ceará (IFCE)
Fortaleza, Ceará, Brazil
Email: renato.xavier@ifce.edu.br

***Abstract*—This paper presents the development and testing of a data acquisition system using the AVR - Internet of Things (IoT) Cellular Mini microcontroller to monitor the temperature of photovoltaic (PV) modules on the roof of the Alternative Energy Laboratory of the Federal University of Ceara (LEA - UFC) in Fortaleza, Brazil. Temperature data are captured by PT100 and DHT11 sensors. Data acquisition is initiated through the connection of each sensor to a dedicated port of the microcontroller, with data transmitted to the cloud using the ThingSpeak platform. The initial phase of the system was tested for one month, providing insights into its operation and the integration of real-time monitoring for renewable energy systems.**

Keywords—IoT; data acquisition; photovoltaics; 5G.

I. INTRODUCTION

As PV systems increase their relevance, real-time data monitoring becomes crucial. For instance, accurate PV modules temperature data play a significant role in determining the efficiency and performance of PV systems [1]. In such context, our paper analyzes the implementation of a PV module temperature data acquisition system using the AVR - IoT Cellular Mini, shown in Figure 1, a microcontroller-based solution that provides real-time monitoring via wireless transmission. The PV modules are installed in a grid connected plant at the LEA - UFC in Fortaleza, Brazil (3° 44' 15" S; 38° 34' 22" W and 21 m). Hence, our proposed system aims to explore the use of cellular IoT technologies to enhance data transmission reliability and overall system scalability.

The mobile network architecture for IoT has been profoundly transformed with the advent of 5G, a technology that offers advanced connectivity, high reliability, and low latency—essential characteristics for the massive integration of IoT devices [2]. In the context of IoT, 5G enables the connection of a vast number of devices, ranging from environmental sensors to remote monitoring systems, with unprecedented efficiency and coverage [3]. The combination of 5G with long-range, low-power cellular communication technologies, such as Long Term Evolution (LTE) for machine communication (LTE-M) and Narrowband IoT (NB-IoT), promotes efficient coexistence of devices in urban and rural areas [4]. These licensed technologies bring benefits in terms of service quality and greater coverage, essential for large-scale and critical IoT applications. LTE-M, part of 3GPP Release 13, operates on licensed frequency bands, integrating seamlessly into 5G network infrastructure [3]. With its ability to provide higher bandwidth and lower latency, LTE-M is particularly useful in scenarios where data transmission demands superior quality of service, such as in PV plant management [4]. The integration of LTE-M into 5G networks becomes essential to ensure efficient coverage in locations where 5G alone cannot reach, such as rural areas. Additionally, LTE-M can coexist with NB-IoT within the 5G architecture, enabling efficiency and the expansion of IoT device use, providing solutions suitable for both large cities and regions with lower population density [3].

The AVR-IoT Cellular Mini microcontroller is a compact and versatile solution for integrating IoT devices into LTE-M networks. This microcontroller enables secure hardware authentication in the cloud using the ATECC608B device, which provides Elliptic Curve Cryptography (ECC), ensuring the integrity and confidentiality of transmitted data [5]. With full support for the Arduino platform, the AVR128DB48 facilitates the development and implementation of customized IoT solutions. The GM02S communication

Figure 1. Microchip AVR-IoT Cellular Mini.

module, which operates with ultra-low power consumption, is responsible for connecting to the LTE-M network, being essential for efficient and secure data transmission. Thus, the integration of the AVR-IoT Cellular Mini with the LTE-M network allows for the creation of robust monitoring systems, such as remote PV plant control, ensuring connectivity and security in the transmission of collected data. The paper is organized as follows: Section I is the introduction, Section II brings out the state of the art, Section III shows the system design and setup, Section IV provides the results and Section V brings the conclusions.

II. State of the Art

The potential of IoT technology and 5G is analyzed in [6], highlighting their main features and the synergy between them. The central proposal is a monitoring system for decentralized PV plants, utilizing an AVR-IoT microcontroller, which stands out for its efficient 5G connectivity and low power consumption. Green IoT emerges as a sustainable approach for efficient communication, data management, and device utilization. Integrating technologies, such as Wireless Sensor Networks (WSN), Cloud Computing (CC), Machine-to-Machine (M2M) Communication, Data Centres (DC) and advanced metering infrastructure, Green IoT reduces energy consumption and promotes environmentally friendly practices across design, manufacturing and usage [7]. This review explores advancements of Green IoT for smart grids, paving the way towards sustainability, covering energy-efficient communication protocols, intelligent energy management, renewable energy integration, demand response, predictive analytics and real-time monitoring. With the advent of cloud computing, the challenges and opportunities resulting from grid expansion can be addressed more efficiently.

The role of energy management systems in both research and industrial practices is discussed in [8], recognizing the importance of both as stakeholders in this domain. The investigation comments on various IoT-related issues concerning PV production, offering insights for future research. 5G use in IoT monitoring systems, such as those utilizing the AVR IoT 5G microchip board, aligns closely with several key aspects of modern 5G applications. First, 5G's role as a General-Purpose Technology (GPT) is highly relevant, enabling the widespread integration of IoT devices into various applications, including monitoring systems.

A critical factor in the efficient use of 5G in IoT systems is network slicing [6]. This concept allows for the logical separation of different bandwidth capacities within the 5G network, ensuring that the Quality of Service (QoS) is maintained across different applications. Network slicing ensures that the devices receive the necessary bandwidth and performance without interference from other network uses. Moreover, the complementarity between physical and virtual components of IoT systems is crucial. The IoT monitoring system integrates the AVR board hardware with the virtual infrastructure provided by 5G networks, such as cloud storage and data processing. This ensures reliable operation between the physical sensors and the software managing the data. Cybersecurity is another vital aspect, ensuring that communication between IoT devices and 5G remains secure. The focus on end-to-end slice security within 5G networks is highly relevant for the project, as it provides protection against cyber threats, ensuring data integrity and privacy in the monitoring system.

Smart cities aim to optimize operations such as waste and traffic management, water supply, crime tracking, and pollution monitoring by leveraging interconnected IoT devices [10]. The challenge lies in the real-time exchange of vast amounts of data required for smart city applications. Device-to-Device (D2D) communications, which offer higher bandwidth and lower latency, are often used to meet this need, as they do not require infrastructure, making them cost- and time-effective. However, the lack of third-party verification in D2D communications introduces security risks. To address this, the paper proposes a secure and lightweight mutual authentication and key agreement protocol for WiFi Direct, utilizing a commit/open pair and the Diffie-Hellman key exchange algorithm. Simulations show that the protocol effectively authenticates D2D devices in the WiFi Direct environment, resisting common attacks such as Denial of Service (DoS) and Man-in-the-Middle (MITM).

The impact of 5G technology in improving connections and increasing efficiency across various fields is discussed in [11]. Operating at higher frequencies, 5G technology offers greater data capacity, lower latency, and higher reliability, enabling the emergence of new services like remote medical procedures and M2M communication. Additionally, the article explores the role of nanomaterials, such as graphene and carbon nanotubes, in the development of nanoantennas, which are essential for enhancing 5G communications.

The implementation of communication networks plays a crucial role in expanding global knowledge [12]. Crowds are utilized to support a large number of internet networks, such as Wi-Fi, or accommodate heterogeneous devices like tablets and smartphones, under various conditions, including downloading or using apps like Spotify. 5G technology, with its ability to connect more users in a specific geographical area, can reduce miscellaneous costs and provide faster solutions. Existing technologies are limited in their capabilities, and society requires greater accuracy and safety in the digital age, which necessitates an upgrade to 5G technology. The Massive IoT (MIoT) concept cannot be achieved with current technologies, as they lack the capacity to handle the massive amounts of information carried by devices and fail to provide the precision required for mission-critical services like industrial automation. The faster and more robust capabilities of data transfer, brought about by upgrades to mobile electronic systems, are accompanied by a wide range of complementary changes. These changes are supported by harmonized spectrum bands, updated international technical standards, new requirements for network operation, innovative cellular devices, and expanded services with a broader array of potential commercial applications.

A systematic outline of the development of 5G-related research until 2020 is presented in [13], based on over

10,000 science and technology publications. The study addresses the emergence, growth, and impact of this area, providing insights into disciplinary distribution, international performance, and historical trends.

5G technology evolution is discussed in [14], highlighting its rapid development, the increase in the number of operators and devices, and the ongoing discussions about innovation, competition, and policy implications. The analysis reveals that 5G technologies complexity has increased over time, resulting in a growing concentration of patents in a few countries, primarily the USA, China, and South Korea. Although Europe has improved its collective position, it still faces significant challenges compared to Asia. The study also notes that the geographical diversity in scientific publications on 5G has increased over time, with a growing participation of universities compared to non-academic organizations. The quality of publications varies depending on the criteria used for evaluation, with China leading in total citation counts, while the USA stands out in terms of average quality. Finally, the article suggests that coordinated action at the European Union level may be crucial for the block to take a leadership role in 5G innovation, emphasizing the need for careful public policies and flexible regulations to maximize the benefits of the technology. The exponential growth of data traffic within modern communication systems has led to a critical shortage in system information rate and increased carbon emissions [7]. Massive multiple-input multiple-output (MIMO) techniques present a promising solution to boost both energy and spectrum efficiency, positioning them as a vital component for next-generation wireless networks.

III. System Design and Setup

The proposed system, shown in Figure 2, is composed of PT100 temperature sensors connected directly to PV modules on the LEA roof. Each sensor is wired to a specific port in the AVR-IoT Cellular Mini microcontroller, which serves as the core data processing unit. Once connected, the microcontroller begins the acquisition of temperature data; sequentially, it sends the data via cellular communication to the cloud-based ThingSpeak platform [15]. To connect the AVR IoT board via 5G, it is necessary to switch the carrier on the pre-installed chip from Truphone to the local carrier that supports 5G. Although the chip on the board supports LTE- M, this configuration enables connectivity with 5G networks. where real-time data visualization and further analysis can be conducted.

ThingSpeak acts as the interface between the microcontroller and the monitoring system, ensuring reliable communication between the hardware and the analytical tools. Data is then stored in a MySQL database, which integrates with ThingSpeak to facilitate data organization and access for extended analysis. During the one-month trial period, the system continuously recorded temperature data from the PV modules, providing an initial dataset to evaluate the system's effectiveness. With the integration of 5G, LEA benefits from faster and more reliable communication, enhancing the efficiency of the entire system. Additionally, one of the most significant advantages of using 5G in the monitoring system is the fast downlink allowing for real-time transmission of large volumes of data. This is critical for PV plants monitoring, enabling immediate detection and response to potential issues, such as overheating or system inefficiencies, in near real-time; this means that the temperature data collected can be analyzed almost instantly.

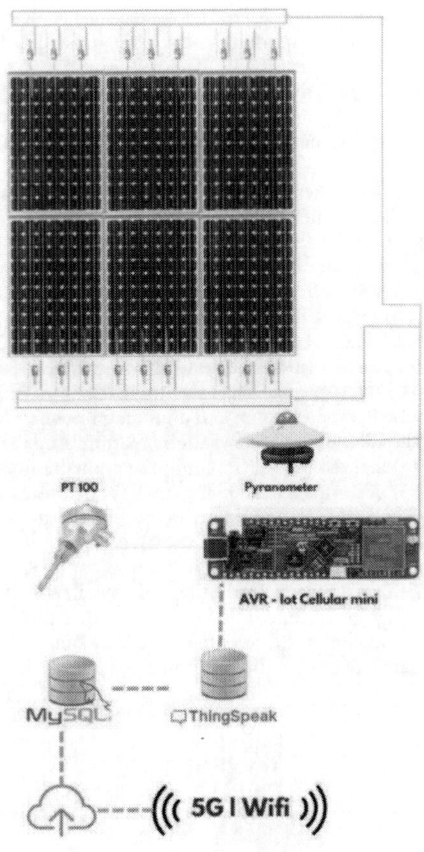

Figure 2. Proposed data acquisition system.

IV. Results

Scalability and device density 5G [8] is designed to handle a much higher density of connected devices compared to previous generations, supporting up to one million devices per square kilometer. Such characteristics are ideal for IoT-based systems like the one in [16], which has a low-cost wireless monitoring system, employing NodeMCU boards, Raspberry Pi, and IoT technologies to monitor and analyze PV modules data. 5G high device density capability allows future expansions of the monitoring system, whether by adding more sensors, integrating additional types of data or scaling the system to larger installations.

Energy efficiency 5G technology is optimized for energy-efficient communication, which is critical for IoT applications like the LEA monitoring system. By reducing device power consumption during data transmission, 5G helps to extend the operational life of the sensors and microcontroller, especially in PV powered systems. This contributes to overall energy savings and ensures that the monitoring system can function effectively without draining excessive power.

Irradiance measurements, recorded every minute, shown in Figure 3, are fundamental for PV systems, allowing an accurate analysis of the amount of solar energy received by the modules. Our study uses an LP02 pyranometer with a sensitivity of 18.56 μV/W/m² to obtain irradiance data throughout a typical dry day, from 00:00 to 23:00 pm. The irradiance peak was recorded at noon, reaching 971.29 W/m²; the measured irradiance reveals a characteristic diurnal pattern, with a gradual increase in the early morning hours, a sharp peak around noon, and a gradual decline until late afternoon. Identifying this pattern is important for predicting PV generation, understanding the contribution of irradiance to PV performance [17]. The analysis of PV actual efficiency is crucial for verifying whether the system is operating as expected or if there are inefficiencies, such as performance loss due to shading, dirt on the panels, or component failures. Such monitoring of irradiance and system efficiency over time allows for the detection of anomalies and prevention of potential failures, contributing to the maintenance and reliability of the PV system.

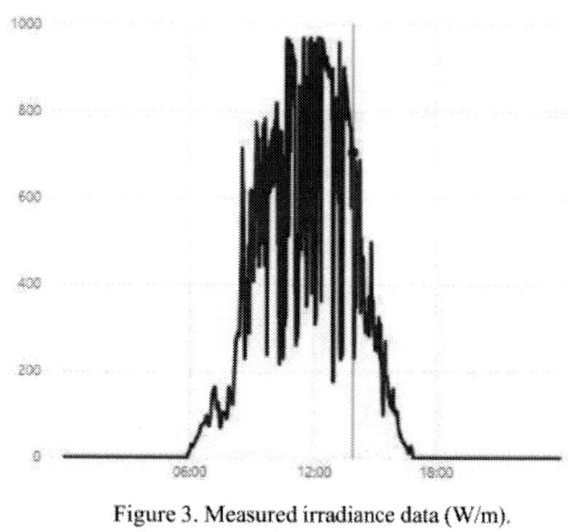

Figure 3. Measured irradiance data (W/m).

Ambient temperature data, shown in Figure 4, is a relevant parameter for evaluating PV performance, as temperature has a negative impact on PV efficiency. The temperatures vary between 22.5°C and 35.03°C, with a concentration between 24°C and 34°C. The data obtained demonstrate stability, with an average temperature of approximately 24.28°C. A PT100 sensor is used to record ambient temperature at regular intervals from 00:00 to 23:00 pm. A characteristic diurnal variation is verified: a gradual temperature increase in the early morning, reaching a peak around noon, coinciding with the time of maximum irradiance, and a gradual decrease through the afternoon. By correlating ambient temperature data with irradiance, it is possible to better understand PV systems behavior under real operating conditions, helping to identify potential cooling needs or system optimization to mitigate the effects of high temperatures, especially in hot climate regions.

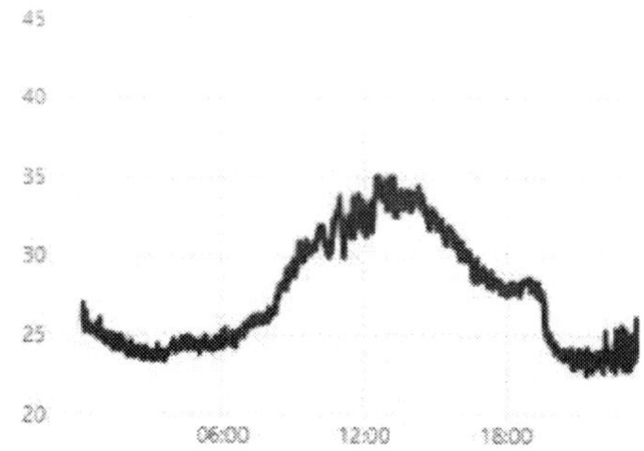

Figure 4. PT100 based measured ambient temperature (°C).

DHT11 based ambient temperature measurements, shown in Figure 5, were done in the same period of the data from the PT100 sensor; the data were recorded and organized for statistical analysis, enabling the identification of patterns and trends. The temperatures vary between 22.94°C and 35.93°C, with a concentration between 23°C and 35°C. The data obtained demonstrate stability, with an average temperature of approximately 24.5°C.

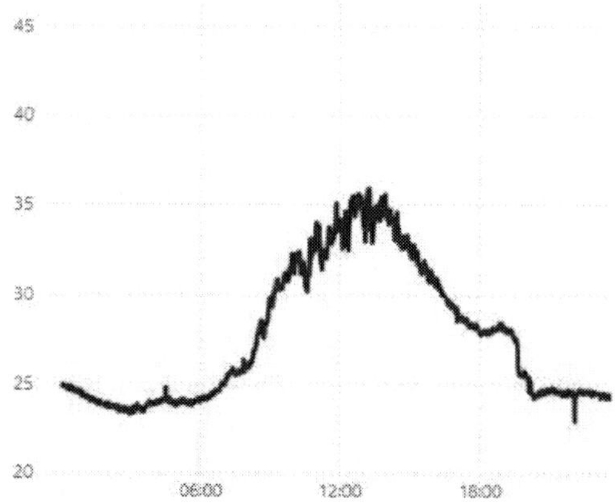

Figure 5. DHT11 based measured ambient temperature (°C).

The 250 Wp module temperature is shown in Figure 6, considering the same day analyzed before. Under intense irradiation, the module surface temperature can increase considerably, often being 20°C to 40°C higher than the ambient temperature. For example, at 8:00 am the module temperature is 31.18°C and the ambient temperature is

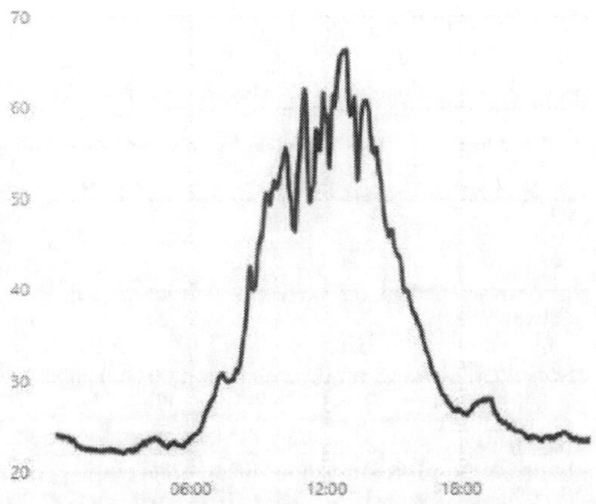

Figure 6. Measured module temperature (°C).

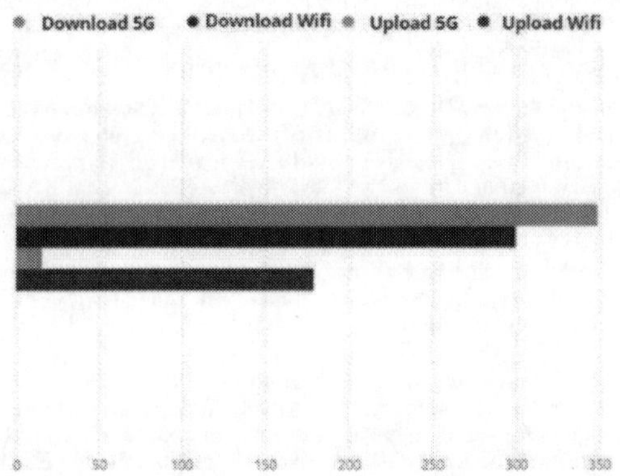

Figure 7. 5G and WiFi download and upload speeds (Mbps).

26.7°C; by 12:00 am, the module temperature rises to 58.43°C, while the ambient temperature reaches 32.47°C. Module temperature data can be used to calculate PV efficiency, as well as to predict electricity production.

The latency, which measures the time necessary for a data packet to travel to its destination and back, is lower for the Wi-Fi connection: WiFi shows a latency of 7 ms, a download speed of 299.4 Mbps, and an upload speed of 178.6 Mbps; 5G shows a latency of 18 ms, a download speed of 346.8 Mbps, and an upload speed of 15.3 Mbps. 5G and WiFi comparative performance is shown in Figure 7.

Comparing PT100 and DHT11 temperature data (see Figure 8), PT100 provides higher accuracy and stability, better reflecting temperature variations throughout the day.

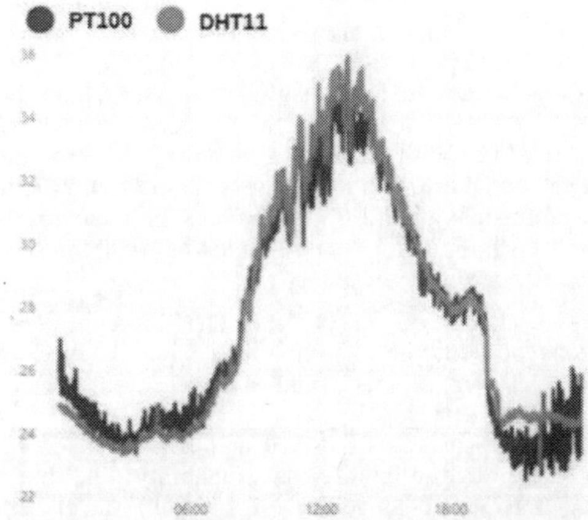

Figure 8. Comparison of DHT11 and PT100 temperature data.

In contrast, DHT11, with lower sensitivity and precision, shows more irregular variations and is less responsive to sudden temperature changes, making the sensor more suitable for general monitoring applications where high accuracy is not critical.

V. CONCLUSIONS

All those mazes are price wise the table pie you reason whyFor the developed data acquisition system, which requires handling and downloading of large amounts of data, 5G proves to be more efficient despite the slightly higher latency. Although Wi-Fi has lower upload and download speeds, its responsiveness is superior, which can be crucial for activities requiring quick response times. On the other hand, 5G connection excels in download speed. With 316.4 Mbps, 5G is ideal for data-intensive activities like high-definition streaming and large file downloads. However, while an upload speed of 31.3 Mbps is acceptable, it falls behind the Wi-Fi performance. The choice between 5G and Wi-Fi may depend on the user's specific needs. For tasks that require high download speeds and can tolerate a little more latency, 5G proves to be advantageous. However, for activities that require low latency and a more stable connection, especially indoors, Wi-Fi may be the better option.

Comparing WiFi and 5G connectivity, our study shows that WiFi provides a lower latency, and a significantly higher upload rate compared to 5G. Such behavior suggests that, while 5G offers advantages in mobility and coverage, WiFi is more suitable for applications that require high data transmission rates and rapid response times, which are essential for continuous, real-time temperature monitoring of PV systems.

LTE-M technology, used in the research, designed specifically for connecting IoT devices, prioritizes low bandwidth and variable latency, characteristics inherent to its proposal for efficient and cost-effective connection for devices with low data throughput demands and high energy savings needs.

Although 5G is known for offering low latency and high bandwidth, LTE-M acts as an enabler for one of the pillars of 5G, massive Machine Type Communication (mMTC). This pillar is aimed at massive communication between IoT devices, which, unlike end-user applications, does not necessarily require high bandwidth or minimum latency. LTE-M has been adapted for the 5G ecosystem, integrating with mMTC and expanding support for the growing demands of the IoT in a massive communication structure. Therefore, it is natural for us to test the data flow in the order of Mbps and the latency can go up to 50 ms according to the standard specifications.

The choice of PT100 and DHT11 considers cost-effectiveness and accessibility for an initial test; however, DHT11 lower accuracy and narrower range can be a limitation. Hence, for future studies, more precise alternatives, such as DS18B20 or other advanced sensors, can be explored to improve data reliability. The first results also motivate a longer testing period, under different meteorological conditions, aiming to assess the long-term reliability and accuracy of the proposed monitoring system. Other areas that need deeper analysis are a) comparison with other monitoring systems for PV applications, including aspects such as accuracy, cost, and ease of implementation, b) integration of additional metrics, such as voltage and current, to offer a broader perspective and c) adaptability of the monitoring system to larger PV plants, including challenges such as communication protocols, power consumption, and data handling.

It is important to mention, no published studies or systems have been identified that employ a similar methodology, making this research innovative in the field, offering a foundation for further exploration and development of the integration of advanced communication technologies with IoT. Future works can expand the area with additional tests, increasing the number of sensors and easing the comparison of experiences.

Acknowledgment

The second author thanks the National Council for Scientific and Technological Development (CNPq) for a researcher scholarship.

References

[1] L. Santos, P. Carvalho and C. Carvalho Filho, "Photovoltaic Cell Operating Temperature Models: A Review of Correlations and Parameters," in IEEE Journal of Photovoltaics, vol. 12, no. 1, pp. 179-190, 10.1109/JPHOTOV.2021.3113156, 2022.

[2] M. Santos, Tecnologia 5G NR e suas aplicações no ramo do Agronegócio (5G NR technology and its applications in the Agribusiness sector). Universidade Estadual Paulista, Guaratinguetá, 2022.

[3] M. Islam, H. Jamil, S. Pranto, R. Das, A. Amin and A. Khan, "Future Industrial Applications: Exploring LPWAN-Driven IoT Protocols". Sensors, 24 (8), 2509. https://doi.org/10.3390/s24082509, 2024.

[4] A. Cavalcante, P. Gomes, M. Marquezini, I. Bonomini and L. Mendes, "Applicability of IoT Technologies for 5G Use Cases in Brazil". Ericsson Research / Ericsson Brazil, Indaiatuba, 2020.

[5] Microchip Technology Inc. AVR-IoT Cellular Mini Hardware User Guide (User Guide DS50003320), 2022.

[6] S. Carvalho, D. Assis, P. Carvalho, L. Ramírez, C. Rios and M. Junior, "5G based data acquisition system applied to photovoltaic plants," in Proc. 22nd Int. Conf. Renewable Energies and Power Quality (ICREPQ'24), Bilbao, Spain, Jun. 2024, pp. 1-6.

[7] T. Abood, I. Hburi and H. Khazaal, "Massive MIMO: An Overview, Recent Challenges, and Future Research Directions," 2021 International Conference on Advance of Sustainable Engineering and its Application (ICASEA), Wasit, Iraq, pp. 43-48, 10.1109/ICASEA53739.2021.9733081, 2021.

[8] C. Rao, S. Sahoo and F. Yanine, "A literature review on an IoT-based intelligent smart energy management systems for PV power generation", Hybrid Advances, vol. 5, 2024, 100136, ISSN2773-207X, https://doi.org/10.1016/j.hybadv.2023.100136.

[9] G. Knieps, "Internet of Things, critical infrastructures, and the governance of cybersecurity in 5G network slicing", Telecommunications Policy, 2024, 102867, ISSN0308-5961, https://doi.org/10.1016/j.telpol.2024.102867.

[10] G. Gaba, G. Kumar, T. Kim, H. Monga and P. Kumar, "Secure Device-to- Device communications for 5G enabled Internet of Things applications". Computer Communications. 169. 10.1016/j.comcom.2021.01.010, 2021.

[11] S. Saraswat, "Why material advancement is necessary approach for the 5G wireless era?" Materials Today: Proceedings, 2023, ISSN 2214-7853, https://doi.org/10.1016/j.matpr.2023.04.435.

[12] B. Gangadhar and K. Sekhar, "Research challenges in 5G communication technology: Study, Materials Today": Proceedings, v o l. 51, Part 1, 2022, Pages 1035-1037, ISSN 2214-7853, https://doi.org/10.1016/j.matpr.2021.07.083.

[13] S. Mendonça, B. Damásio, L. Freitas, L. Oliveira, M. Cichy and A. Nicita, "The rise of 5G technologies and systems: A quantitative analysis of knowledge production", Telecommunications Policy, vol. 46, Issue 4, 2022, 102327,ISSN 0308-5961, https://doi.org/10.1016/j.telpol.2022.102327.

[14] P. Parcu, M. Rossi and T. Brennan, "Innovation in 5G technology: Leadership, competition, and policy issues". Telecommunications Policy. 46. 102349. 10.1016/j.telpol.2022.102349, 2022.

[15] https://thingspeak.mathworks.com/

[16] M. Radia, M. Nimr and A. Atlam, "IoT-based wireless data acquisition and control system for photovoltaic module performance analysis", e-Prime - Advances in Electrical Engineering, Electronics and Energy, vol. 6, 2023, 100348, https://doi.org/10.1016/j.prime.2023.100348.

[17] T. Carneiro, P. Carvalho, H. Santos, M. Lima and A. Braga, "Review on photovoltaic power and solar resource forecasting: Current status and trends", J. Sol. Energy Eng. 144 (1), 2022, 010801. https://doi.org/10.1115/1.4051652.

On Zero-shot Learning in Neural State Estimation of Power Distribution Systems

Aleksandr Berezin*, Stephan Balduin*, Eric MSP Veith*,
Thomas Oberließen† and Sebastian Peter†
*OFFIS – Institute for Information Technology, Oldenburg, Germany
e-mail: {name.surname}@offis.de
†Institute of Energy Systems, Energy Efficiency and Energy Economics
Technical University of Dortmund, Germany

Abstract—**This paper addresses the challenge of neural state estimation in power distribution systems. We identified a research gap in the current state of the art, which lies in the inability of models to adapt to changes in the power grid, such as loss of sensors and branch switching, in a zero-shot fashion. Based on the literature, we identified graph neural networks as the most promising class of models for this use case. Our experiments confirm their robustness to some grid changes and also show that a deeper network does not always perform better. We propose data augmentations to improve performance and conduct a comprehensive grid search of different model configurations for common zero-shot learning scenarios.**

Keywords-neural state estimation; zero-shot learning; transfer learning; graph neural networks.

I. INTRODUCTION

Power System State Estimation (PSSE) is the task of inferring the "state" of an electrical power grid from real-time data collected by various sensors distributed across the system. The "state" in this context generally refers to the voltage magnitudes and phase angles at each bus in the grid.

For many years, PSSE was mainly performed for the transmission grids using simplifying assumptions such as near-DC power flow and computational methods with poor scalability [1]. This is enabled by balanced operation with a relatively simple, predominantly linear topology of transmission grids, given their scale and structure.

On the contrary, distribution grids, which transport electricity from substations to end consumers, present distinct challenges. Their unbalanced nature, radial or weakly meshed topology, high R/X ratios, and cost inefficiency to achieve sufficient sensor coverage complicate the state estimation process. Initially designed with transmission systems in mind, conventional methods often struggle to provide accurate state estimation in these more complex, dynamic, and less predictable distribution systems [1].

However, with the proliferation of Distributed Energy Resources (DERs) and other complex consumers, grid operators are facing the necessity of performing PSSE for distribution grids. Additionally, §14a of the German Energy Industry Act effectively requires operators to develop transparency in distribution grids in order to align consumption with production from renewable energy sources, which requires PSSE.

In this paper, we begin by reviewing relevant prior work in Section II, followed by a formal statement of our research question in Section III. Section IV details the methodology, including model selection, data preprocessing, and experimental setup. We present and analyze our results in Section V and discuss their implications. Finally, Section VII summarizes our findings and suggests directions for future work.

II. RELATED WORK

The traditional and most widely-used approach for PSSE is the Weighted Least Squares (WLS) method [2]. This algorithm minimizes the sum of the squared differences between the observed and estimated measurements, with each term being weighed inversely proportionally to the square of the measurement error standard deviation.

However, the WLS algorithm is computationally intensive. Its time complexity is generally considered to be $\mathcal{O}(N^3)$ in the number of buses N, assuming a dense system matrix [2]. This is due to the need for matrix inversions and solving linear equations. This complexity can become a limitation for large-scale power systems with thousands of buses, leading to significant computational burden and time constraints, especially when real-time or near-real-time estimations are required. Additionally, WLS assumes that all error distributions are Gaussian, a condition that may not always hold true in practice.

To overcome these limitations, an increasing number of publications instead use Artificial Neural Networks (ANNs) for PSSE, a combination that is called Neural State Estimation (NSE). ANNs may be able to perform the calculation faster than iterative solutions and achieve a higher solution quality simultaneously [3][4]. However, like all Machine Learning (ML) methods, the performance of ANNs is contingent on the quality and quantity of the available training data. Therefore, NSE approaches are usually valid only for the grid they have been trained on. Once the topology or characteristics of nodes change, the ANN needs to be retrained. This is known as the problem of Transfer Learning (TL).

The most logical way to overcome this limitation is to use models that incorporate information about the graph topology into their calculations. Such models are known under an umbrella term Graph Neural Networks (GNNs). Expectedly, recent years have seen a high volume of publications that propose utilizing GNNs for NSE in various ways. To name a few examples:

- Park *et al.* [5] lays important groundwork in comparing different matrix representations of graphs within the Graph Convolutional Network (GCN) model;

- Kundacina *et al.* [6] utilizes Graph Attention Networks (GATs) with a different graph representation of the power grid;
- Hossain and Rahnamay-Naeini [7] explore the possibility of utilizing temporal correlations in the datasets using recurrent GCNs.

However, to our knowledge, none of these research projects specifically considered the problem of Zero-Shot Learning (ZSL) in PSSE. The contribution of this work is in setting up multiple evaluation scenarios for ZSL and testing different configurations of GNNs in them.

III. Research question

When discussing the ability of a model to generalize to different grid topologies, it is important to differentiate between *homogeneous* and *heterogeneous* modes of TL. In general, homogeneous TL mode means that the source and target data are in the same feature space, while in heterogeneous TL mode, they are represented in different feature spaces.

In the context of power grids, this is the difference between two use cases. In the homogeneous case, the power grid remains the same, but some connections between its nodes appear or disappear due to changes in switch states or elements going in and out of service. In the heterogeneous case, the model trained on one grid is used to make predictions about a completely different grid [8].

This distinction becomes very important in production environments. Integrating a model into the control system of a real grid naturally takes time, and training the model on that specific grid could be incorporated into this process without noticeably slowing it down. On the other hand, changes in grid topology due to switching can happen suddenly and unpredictably, and the model must adapt to them in real-time.

There is also another way in which the data distribution can shift in the context of PSSE: the observable subset of buses can change, which changes the amount of input data points available to the model. This can also be considered a form of homogeneous TL.

A subset of TL is Zero-Shot Learning (ZSL). This scenario excludes the possibility of fine-tuning the model on the new distribution and evaluates its performance directly after the transfer. In this project, we specifically focus on ZSL because it is more representative of real-life situations where a model must make predictions immediately after a topology change without access to any training data for fine-tuning. In other words, the model should be *robust* to distributional shifts.

Of course, in practice, a model can be fine-tuned to provide the best performance for the new topology. Still, until this process is complete, the previous version of the model has to substitute for it and provide good enough estimations, even if they are of lower quality.

The research question for this paper is which existing models in application to the PSSE problem are robust to changes in the data distribution, specifically:

A To the reduction of the subset of observable buses;

B To grid topology changes resulting from changing switch states;

C To transfer to a completely different power grid.

IV. Methodology

A. Model selection

The general question of model selection for NSE was addressed by us previously in [9]. The main conclusion from that paper was the selection of GNNs as the most promising direction for further research. Now, we will perform a similar comparison study within the GNN family. We are comparing four models using the implementations provided by PyTorch Geometric framework [10]:

1) Graph Convolutional Network (GCN) as proposed in [11]
2) Graph Attention Network (GAT) as proposed in [12]
3) Graph Isomorphism Network (GIN) as proposed in [13]
4) Graph Sample and Aggregate (GraphSAGE) as proposed in [14]

B. Graph representation of power systems

A successful application of GNN models naturally depends on how well the underlying data can be represented in the graph format. The first step is to represent buses in the grid as nodes of the graph and lines as its edges. In this project, we also represented transformers as edges without any additional parameters. For this to work, the voltage levels across the transformer must be normalized to avoid large voltage gradients.

It is also theoretically advantageous to use a weighted graph with line admittances as weights. Admittances are chosen because the graph Laplacian operator assumes higher edge weights to mean a higher correlation between nodes. This operator is, in turn, used in both the GNN models and the feature propagation algorithm discussed in the next subsection. It should be noted that the models in question support neither complex-valued weights nor multidimensional weights, so we have to use the magnitude of the true complex impedance.

However, using admittance instead of impedance as edge weights becomes a problem for representing closed switches, which have zero impedance and, therefore, infinite admittance. This problem is solved by fusing buses connected by closed bus-to-bus switches into one bus. This is complicated because multiple closed switches are often connected to the same bus, so a naive approach of fusing adjacent buses in random order does not work. Instead, we use an iterative algorithm. Firstly, we build an auxiliary graph of just the closed bus-to-bus switches with buses as nodes and switches as edges. In this graph, nodes with a degree of one can be safely removed (fused with their adjacent buses). This will, in turn, lower the degree of the adjacent node. Eventually, every node will reach a degree of one and can be fused until every connected component of the auxiliary graph is fused into a single node.

C. Data preprocessing

GNN models are geometrically isodimensional, meaning that each output node must have a corresponding input node.

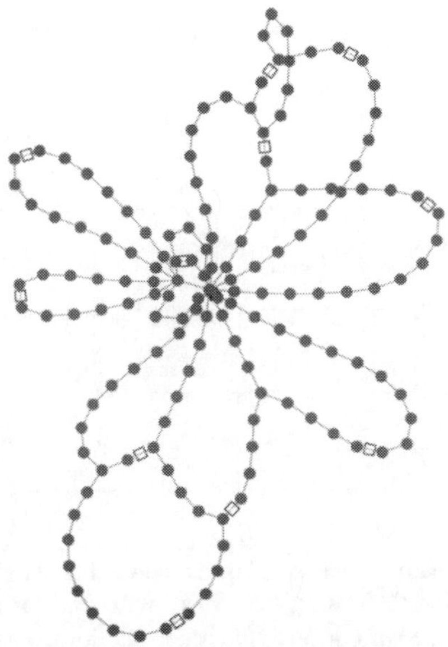

Figure 1: A visualization of the SimBench 1-MV-urban–1-sw grid

This presents a problem for the PSSE use case, where we lack input features for many, if not most, input nodes. The question, therefore, is: How do we initialize the missing features in the input data?

The solution we chose is the feature propagation algorithm from [15], which interpolates missing node-level features by solving a heat equation with known features as boundary conditions. This results in a smooth interpolation of features between known nodes and forms a starting point for the subsequent application of GNNs.

D. Datasets

The main dataset used in this project is the SimBench *1-MV-urban–1-sw*, a 147-node, 10 kV medium voltage grid [16] depicted in Figure 1. It is composed of a grid model and a per-bus complex (active and reactive) power yearly time series. To calculate the resulting grid state, we performed a power flow calculation using SIMONA energy system simulation [17]. The resulting dataset comprises the base data and a year of complex voltage time series with a 15-minute temporal resolution. This dataset is hereafter called PQ.

Most grid branches in this model are of the open loop type, which means an open switch (depicted as a square) connects two separate branches. To simulate a realistic topology change, we made a line in one of the open loop branches inoperable, resembling a line fault, and closed the loop switch to resupply all nodes. Performing this operation on different branches resulted in multiple variations of the base grid topology. Afterward, we reran the simulation for each variation to obtain a topology change dataset, which is referenced hereafter as TC.

Unfortunately, the base dataset did not contain information about measurement devices. Therefore, we had to choose observable nodes randomly based on an observability level of 50%, which we assume is realistic for distribution grids. This means that the state estimator has access to true voltage values for half of the grid buses.

An auxiliary dataset used in the heterogeneous ZSL experiments is based on the CIGRE medium voltage distribution network from Pandapower [18]. It is a much smaller grid with only 15 nodes, which allows us to study how the complexity of the grids affects the performance of ZSL. The voltage data for it is generated using the Midas simulation framework [19]. The shorthand name for this dataset is MV.

E. Use cases

Our experiments will be composed of three benchmarks that we call use cases. They correspond to the three subquestions of the main Research question.

In the first use case corresponding to subquestion A, we train the model on the grid with a baseline level of observability and then linearly reduce it from the baseline level to zero at testing time. Of course, the model performance decreases along with this reduction. The shorthand name of this use case is observability degradation (OD).

The second use case corresponds to subquestion B and tests ZSL for homogeneous topology changes. In it, we split the TC dataset in a 50:50 ratio, train the model on the first part, and evaluate on the second. We also evaluate another model trained on the PQ dataset on the TC testing subset to see if the model needs to observe the topology changes happening in order to be able to adapt to them at testing time, but our null hypothesis is that this is not the case. This scenario has the shorthand name TC1, and the former, where the model is trained on the TC dataset, is called TC2.

The third use case corresponds to subquestion C and covers the heterogeneous ZSL scenario. Here, we transfer the model between the PQ and MV datasets in both directions, that is, training on one and then testing on another. The scenario in which the model is trained on PQ and tested on MV has the shorthand name PQ2MV, and the other has MV2PQ.

F. Experiment setup

Let us now establish the full hyperparameter space for the models in question. It consists of the following dimensions:

- Model, as listed in the Model selection subsection. Categorical parameter with four values.
- Number of layers in the model. Integer parameter that we limit to 10.
- Use of feature propagation (as opposed to initializing the missing features with zeros). Boolean parameter.
- Use of admittances as edge weights (as opposed to not using any edge weights). Boolean parameter.

Unfortunately, preliminary experiments have demonstrated that the hyperparameter space is not separable, meaning that a full grid search of the space is required. We performed this search for all model configurations and use cases and collected the Mean Squared Error (MSE) metric for each one. The results of this experiment are organized into an evaluation table where rows correspond to model configurations and columns are the following:

TABLE I: BEST CONFIGURATIONS FOR OBSERVABILITY DEGRADATION (OD)

model	layers	fp	adm	mse
GraphSAGE	3	True	False	0.86
GraphSAGE	2	True	True	0.87
GraphSAGE	3	True	True	0.87
GraphSAGE	2	True	False	0.88
GCN	3	True	False	0.90

TABLE II: BEST CONFIGURATIONS FOR SWITCHING CHANGES (TC1)

model	layers	fp	adm	mse
GraphSAGE	3	True	False	0.31
GraphSAGE	1	True	True	0.33
GAT	1	True	False	0.33
GraphSAGE	2	True	False	0.34
GraphSAGE	3	True	True	0.34

TABLE III: BEST CONFIGURATIONS FOR SWITCHING CHANGES (TC2)

model	layers	fp	adm	mse
GraphSAGE	1	True	False	0.32
GraphSAGE	1	True	True	0.32
GAT	3	True	False	0.32
GAT	1	True	True	0.32
GAT	1	True	False	0.32

TABLE IV: BEST CONFIGURATIONS FOR HETEROGENEOUS TRANSFER (PQ2MV)

model	layers	fp	adm	mse
GCN	1	False	True	0.30
GCN	2	False	True	0.87
GCN	4	False	True	0.97
GAT	2	False	False	1.06
GraphSAGE	2	False	True	1.26

1) "model" is the name of the model;
2) "layers" is the number of layers in the model;
3) "fp" is a binary parameter indicating whether feature propagation is used;
4) "adm" is a binary parameter indicating whether admittance weights are used;
5) "mse" is the value of MSE for the configuration defined by the above parameters.

The full table is available in our repository in the "results.csv" file, and in the next section, we will use subsets of it as illustrations of results.

V. EVALUATION

In this section, we will analyze the results of the full grid search, attempting to answer the following questions:

1) Which model configurations perform best for each use case?
2) How does model complexity affect performance?
3) How do the data augmentations proposed in the Methodology section affect performance?
4) How is performance on different tasks correlated?

The answers to these questions will then be used to answer the main research questions from Section III.

A. Ranking model configurations

To interpret the numerical results listed in this section, it is useful to keep in mind the baseline value of MSE obtained by evaluating the models trained on the first half of the PQ dataset on the second half of the same dataset and taking the best result. This value is 0.32. We can then broadly say that ZSL is possible in scenarios where the value of MSE after the topology change does not significantly exceed it.

The winning model for the first use case is Graph Sample and Aggregate (GraphSAGE) utilizing feature propagation. It also appears from Table I that there is a sweet spot in model complexity of 2-3 layers.

In both scenarios of the second use case (Tables II and III), the model rating is similar, which suggests that the tasks themselves are similar as well. The winning models are GraphSAGE and GAT, also with the help of feature propagation and, curiously, in their shallowest versions, with single-layer models showing some of the best results. The main observation, however, is that the MSE values are identical to the baseline, meaning that homogeneous ZSL works very well.

In the third use case, we see a significant difference between the two scenarios. In the first scenario (Table IV), where the model is transferred from a larger to a smaller grid, the best-performing by a large margin is a single-layer GCN, which is the simplest of all the compared models. A possible explanation is that there are only a few correlations that are reusable between grids, which the simple model can capture. Any more complex model picks up too many correlations that are specific to the grid it was trained on and then misapplies them. In the second scenario (Table V), the results are mixed between complex and simple models, and we cannot come up with a sound theoretical interpretation of this result.

B. Impact of model complexity

In Figure 2, we plot the performance of models against their trainable parameter counts. We also plot the average for all models of a given complexity: the "Mean" line on the graph.

Here, we can see a break point at about 84 parameters or 8 layers, starting from which the performance of models becomes much more consistent between use cases and configurations. The explanation for this effect is the over-smoothing phenomenon described in [20]. In short, GNN layers of all

TABLE V: BEST CONFIGURATIONS FOR HETEROGENEOUS TRANSFER (MV2PQ)

model	layers	fp	adm	mse
GAT	6	True	False	0.62
GAT	2	True	False	0.64
GCN	2	True	True	0.67
GIN	1	True	True	0.69
GAT	2	False	True	0.78

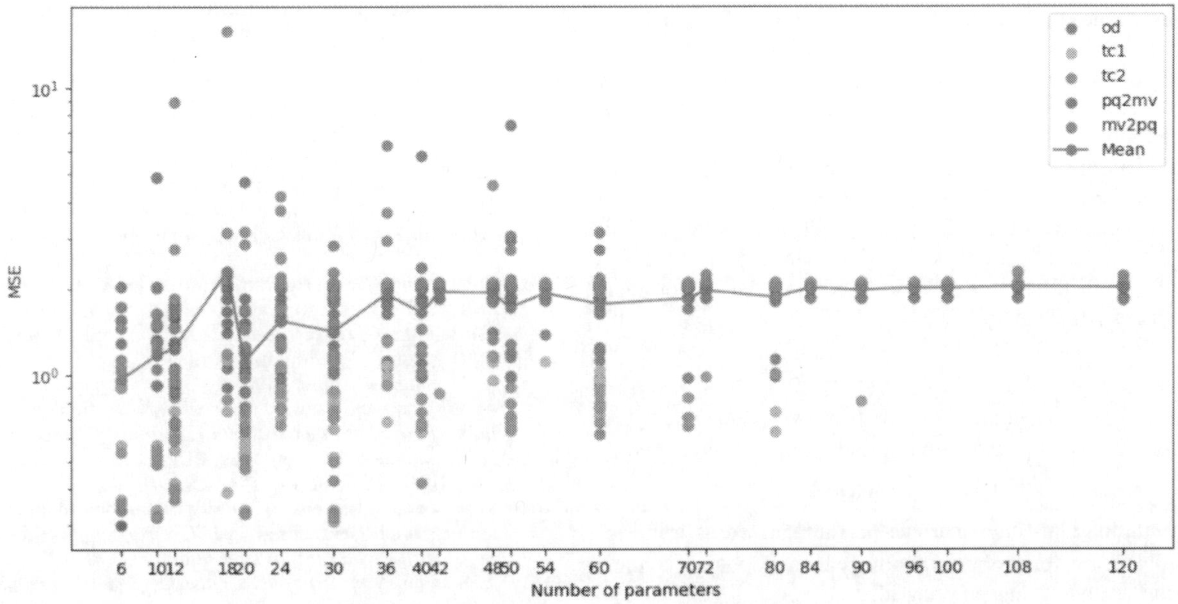

Figure 2: Performance as a function of model complexity

TABLE VI: IMPACT OF DATA AUGMENTATIONS

fp	adm	od	tc1	tc2	pq2mv	mv2pq
True	True	1.33	1.17	1.13	2.19	1.80
True	False	1.25	1.03	1.06	1.99	2.55
False	True	1.51	1.24	1.25	1.93	2.00
False	False	1.49	1.23	1.10	1.98	2.09

architectures tend to act as low-pass filters, which effectively averages the output values over multiple iterations. Eventually, the model converges to an output where the values at all nodes of the graph are identical.

Since we are not interested in over-smoothed results, we can drop the model configurations that output them from further analysis. Therefore, the following sections will use the results table truncated to a maximum of 7 layers.

C. Impact of data augmentations

In Table VI, we average the performance values across all model configurations, leaving only the data augmentations as parameters. The results are unsurprising and follow the observations we already made previously. Homogeneous scenarios benefit from feature propagation but are held back by admittance weights. In the heterogeneous scenarios, we once again see a split where MV2PQ benefits from admittance weights and PQ2MV does not. On average, the impact of the augmentations is not very significant.

D. Correlation analysis

To explore the correlations between hyperparameters and use cases, we compute the Pearson correlation matrix between their associated performance values in Figure 3. Here, we also use the number of trainable parameters ("#params") instead of layers to compare model complexity more fairly. Note that since lower MSE means better performance, a positive correlation between a hyperparameter and performance is shown as negative and vice versa.

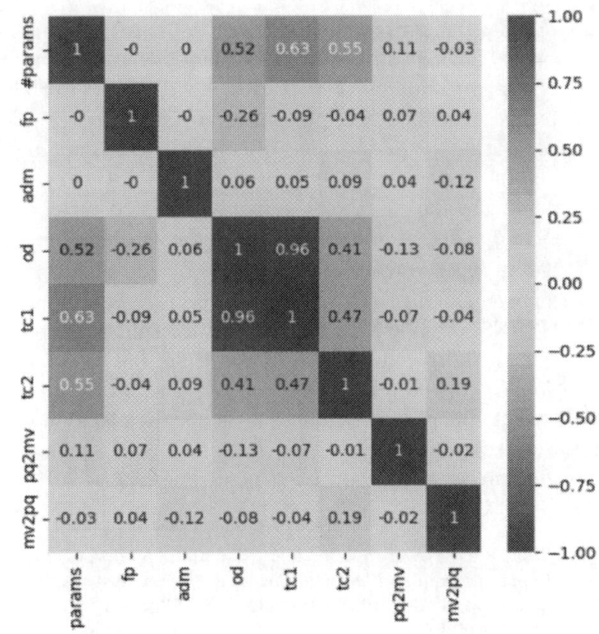

Figure 3: Performance correlation between use cases

The analysis confirms the conclusions that we made previously: model complexity is detrimental to performance for all use cases except MV2PQ, and our proposed augmentations only marginally affect performance, with feature propagation being most useful in homogeneous scenarios.

VI. Limitations and Future Work

During the evaluation stage of this research, it became evident that MSE alone does not convey enough information to confidently make conclusions about ZSL performance of the models. However, we could not find a better alternative in the literature.

The main problem is that MSE only shows us the instantaneous performance and does not account for the transfer process. An ideal metric M for ZSL and TL experiments would be a differential one that takes into account the magnitude of the change in the underlying grid ΔG and the performance change ΔP, for example,

$$M = \frac{\Delta P}{\Delta G}$$

However, an algorithm to compute ΔG is not trivial to develop. We hope to tackle this problem in our future work.

VII. Conclusion

The findings of this paper can be summarized as follows:

1) GNNs are very robust to homogeneous topology changes in the underlying power grid.
2) Some GNNs can perform well in a zero-shot transfer from a larger grid to a smaller one, but not in the other direction.
3) Despite the conventional wisdom in the ML community being "Scale is all you need," scaling GNNs up is not the best way to improve NSE performance. This is explained by the oversmoothing phenomenon [20].
4) Measuring the performance of NSE by MSE is not always helpful, especially in the context of ZSL. However, we could not find another commonly accepted metric. The development of such a metric appears to be a research gap at the moment.

Acknowledgments: This research is a part of project TRANSENSE, funded by the German Federal Ministry for Economic Affairs and Climate Action (FKZ 03EI6044A).

Availability of data and source code: The source code for this project, along with the datasets, is openly available in the following repository:
https://gitlab.com/transense/nse-tl-paper/tree/IARIA

References

[1] F. F. Wu, "Power system state estimation: A survey," *International Journal of Electrical Power & Energy Systems*, vol. 12, no. 2, pp. 80–87, 1990, ISSN: 0142-0615. DOI: 10.1016/0142-0615(90)90003-T.

[2] A. Abur and A. G. Expósito, *Power System State Estimation*. CRC Press, Mar. 2004, ISBN: 9780203913673. DOI: 10.1201/9780203913673.

[3] K. R. Mestav, J. Luengo-Rozas, and L. Tong, "State estimation for unobservable distribution systems via deep neural networks," in *2018 IEEE Power & Energy Society General Meeting (PESGM)*, 2018, pp. 1–5. DOI: 10.1109/PESGM.2018.8586649.

[4] S. Balduin, T. Westermann, and E. Puiutta, *Evaluating different machine learning techniques as surrogate for low voltage grids*, Oct. 2020. DOI: 10.1186/s42162-020-00127-3.

[5] S. Park, F. Gama, J. Lavaei, and S. Sojoudi, "Distributed power system state estimation using graph convolutional neural networks," in *Hawaii International Conference on System Sciences*, 2023.

[6] O. Kundacina, M. Cosovic, D. Miskovic, and D. Vukobratovic, "Distributed nonlinear state estimation in electric power systems using graph neural networks," pp. 8–13, 2022. DOI: 10.1109/SmartGridComm52983.2022.9960967.

[7] M. J. Hossain and M. Rahnamay-Naeini, "State estimation in smart grids using temporal graph convolution networks," in *2021 North American Power Symposium (NAPS)*, IEEE, Nov. 2021, pp. 01–05. DOI: 10.1109/naps52732.2021.9654642.

[8] S.-G. Yang, B. J. Kim, S.-W. Son, and H. Kim, "Power-grid stability predictions using transferable machine learning.," *Chaos*, vol. 31 12, p. 123 127, 2021. DOI: 10.1063/5.0058001.

[9] A. Berezin, S. Balduin, T. Oberließen, E. Veith, S. Peter, and S. Lehnhoff, "Application of recurrent graph convolutional networks to the neural state estimation problem," *International Journal of Electrical and Electronic Engineering & Telecommunications*, pp. 209–215, 2023, ISSN: 2319-2518. DOI: 10.18178/ijeetc.12.3.209-215.

[10] M. Fey and J. E. Lenssen, "Fast graph representation learning with PyTorch Geometric," in *ICLR Workshop on Representation Learning on Graphs and Manifolds*, 2019.

[11] T. Siameh, *Semi-supervised classification with graph convolutional networks*, Dec. 2023. DOI: 10.13140/RG.2.2.22993.71526.

[12] P. Veličković, G. Cucurull, A. Casanova, A. Romero, P. Liò, and Y. Bengio, *Graph attention networks*, 2018.

[13] K. Xu, W. Hu, J. Leskovec, and S. Jegelka, *How powerful are graph neural networks?* 2019.

[14] W. L. Hamilton, R. Ying, and J. Leskovec, *Inductive representation learning on large graphs*, Long Beach, California, USA, 2017.

[15] E. Rossi, H. Kenlay, M. I. Gorinova, B. P. Chamberlain, X. Dong, and M. M. Bronstein, "On the unreasonable effectiveness of feature propagation in learning on graphs with missing node features," Proceedings of Machine Learning Research, vol. 198, B. Rieck and R. Pascanu, Eds., 11:1–11:16, Dec. 2022.

[16] S. Meinecke, D. Sarajlić, S. R. Drauz, A. Klettke, L.-P. Lauven, C. Rehtanz, *et al.*, "Simbench—a benchmark dataset of electric power systems to compare innovative solutions based on power flow analysis," *Energies*, vol. 13, no. 12, p. 3290, Jun. 2020, ISSN: 1996-1073. DOI: 10.3390/en13123290.

[17] J. Hiry, "Agent-based discrete-event simulation environment for electric power distribution system analysis," en, Ph.D. dissertation, 2021. DOI: 10.17877/DE290R-22549.

[18] L. Thurner, A. Scheidler, F. Schäfer, J.-H. Menke, J. Dollichon, F. Meier, *et al.*, "Pandapower—an open-source python tool for convenient modeling, analysis, and optimization of electric power systems," *IEEE Transactions on Power Systems*, vol. 33, no. 6, pp. 6510–6521, 2018. DOI: 10.1109/TPWRS.2018.2829021.

[19] S. Balduin, E. Veith, and S. Lehnhoff, "Midas: An open-source framework for simulation-based analysis of energy systems," in *Simulation and Modeling Methodologies, Technologies and Applications*. Springer International Publishing, 2023, vol. 780, pp. 177–194, ISBN: 978-3-031-43823-3. DOI: 10.1007/978-3-031-43824-0_10.

[20] K. Oono and T. Suzuki, "Graph neural networks exponentially lose expressive power for node classification," in *International Conference on Learning Representations*, 2020.

Federated Learning for Distributed Load Forecasting: Addressing Data Imbalance in Smart Grids

Alexander Wallis*, Sascha Hauke**, Hannah Jörg**, Konstantin Ziegler*
*Department of Computer Science
**Department of Interdisciplinary Studies
University of Applied Sciences Landshut
Landshut, Germany
e-mail: {alexander.wallis | sascha.hauke | hannah.joerg | konstantin.ziegler}@haw-landshut.de

***Abstract*—The integration of renewable energy resources transforms traditional energy systems, introducing prosumers – entities that both produce and consume energy – as key participants in modern Smart Grids. Effective load forecasting is mandatory for optimizing energy resources and grid stability. Federated Learning (FL) has emerged as a promising approach for distributed training of Machine Learning (ML)-based forecasting models. This enables collaborative model optimization across multiple prosumers while preserving data privacy. However, the impact of unbalanced data sets across participants remains a critical challenge in terms of potentially effecting learning convergence and forecast accuracy. In this work, we define and implement a FL system based on real-world electricity consumption data from a variety of prosumers. Experimental results demonstrate the trade-off between centralized and federated learning approaches, providing insights into addressing data heterogeneity in FL systems. These insights highlight the potential of FL to support the evolution of distributed energy systems while ensuring data-privacy and scalability. Future research directions include other strategies to migrate the effect of data imbalances and further improve the efficiency of federated optimization for dynamic energy systems.**

Keywords-Short-Term Load Forecasting; Federated Learning; Smart Grid; Data Privacy; Distributed Data.

I. INTRODUCTION

Accurate load forecasting is mandatory for stable and reliable Smart Grid (SG) operation. But, the accuracy of load forecasting models, in particular ML based models, highly depends on the amount and quality of available training data [1]. Especially on smaller grid levels, e.g., low-voltage grids, or even residential household levels, the available electricity consumption data are very limited. But, with the rise of *prosumers* – consumers also able to produce electricity – prediction models on exactly this grid level is crucial for network management tasks [2].

Even if households are able to record and transmit electricity consumption data through smart meter utilization, the grid operator needs sufficient data storage and computational resources to process the data. Otherwise, the gathered data must be transferred for further processing. This transfer rises data privacy concerns and is even prohibited by law, e.g., General Data Protection Regulation [3]. The ability of information and behavior retrieval based on leakage of electricity consumption data has already been shown in the past [4]–[6].

Here, FL seems to be a promising approach to develop a single ML model for electricity consumption forecasting with distributed data sets – and at the same time satisfying data privacy regulation [7]. In contrast to the traditional approach, where the training of ML model is done centralized, this task is shifted to each user individually.

In [8], FL was first used by McMahan *et al.* to train prediction models on mobile devices through users' keyboard inputs. Afterwards, applications with FL were proposed in various fields, e.g., medical and health care, industrial engineering, finance, transportation [9]–[11].

For SG development, various FL approaches were proposed, too. In [12], FL is used for anomaly detection in terms of energy usage with a detection rate compared to centralized approaches. The authors in [13] present a conceptual framework for secure FL usage in SG environments with focus on vertical and horizontal data distribution over the clients. A detailed overview of further interesting FL researches in the field of SGs is given in [14].

Although, FL can be a promising approach for distributed load forecasting, the impact of unbalanced data sets among the clients is unclear. To evaluate FL in the context of prosumer-level load forecasting, we present the following contributions in this work:

- Definition and implementation of FL system composed of a variety of prosumer based on real-world electricity consumption data.
- Comparison of forecast accuracy between a centralized and a federated learning approach for model optimization.
- Investigation of the influence of unbalanced data sets within a federation on the learning convergence and the overall forecasting error.

This work is organized as following. First, the necessary background information as well as notation and terminologies are given in Section II. Second, the proposed FL approach is described in detail and the different experiments conducted are described in Section III. Third, the experiment results are presented, compared, and subsequently evaluated and discussed w.r.t. forecasting accuracy in Section IV. Fourth and last, the insights gained from the experiments' results are summarized and starting points for further research are given in Section V.

II. Background

Before further detailing the conducted experiments in Section III, we give the respective problem formulation (Section II-A) and background information on FL (Section II-B) as well as an overview of related work (Section II-C).

A. Problem Formulation

Basically, the load forecasting problem can be categorized into three groups based on the forecast horizon: (i) short-term, (ii) middle-term and (iii) long-term load forecasting. In this work, attention is paid on Short-Term Load Forecasting (STLF), since we are interested in a household's next day electricity consumption.

Let $\mathbf{x}_d = (x_d^{(0)}, ..., x_d^{(T)}) \in R^T$ be a household's consumption of day d divided into T time intervals. Further, let $\mathbf{y}_d = (y_{d+1}^{(0)}, ..., y_{d+1}^{(T)}) \in R^T$ be the next day's electricity consumption, then $\mathcal{D} = \{(\mathbf{x}_i, \mathbf{y}_i) | i = 0, ..., D\}$ is the data set composed of input-output pairs for a total of D days. Now, a supervised learning approach approximates a function $\mathbf{y}_d \approx \hat{f}(\mathbf{x}_d)$ for the following optimization problem:

$$\arg\min_{\hat{f} \in \mathcal{H}} \frac{1}{n} \sum_{i=1}^{n} L(\hat{f}(\mathbf{x}_i), \mathbf{y}_i) \quad (1)$$

where $L(\cdot)$ is the desired cost function to be minimized.

Typically, in a centralized learning setting, this is done by collecting each household's data and subsequently by training a combined forecasting model, which is afterwards distributed to every household. Indeed, this rises all of the problems and concerns described earlier (see Section I) and FL is a promising approach to tackle all of them.

B. Federated Learning

Contrary to the centralized learning, a FL approach guarantees data-privacy by preserving prosumers' consumption data locally. A collaboration of prosumer – a so-called *federation* – trains a STLF model by only exchanging respective model parameters. Typically, the participants within a federation are called clients but in this work the terms clients, prosumers and households are used interchangeably. Let $\mathcal{P} = \{p^{(i)} | i = 0, ..., N\}$ be the set of N prosumers then the FL procedure involves the following steps:

1) **Distribution** of the initial global model to all prosumers which are part of the federation $p \in \mathcal{P}$.
2) **Training** of the global model by adjusting it's parameters based on the local data set of every prosumer.
3) **Returning** the adjusted model parameters to a central unit, e.g., trusted 3rd party, data center, one of the participants.
4) **Aggregation** of all received parameters by a predefined `aggregate`-function and integration into the global model.

This whole procedure, also depicted in Figure 1, is repeated over a defined number of *communication rounds* r. Interestingly, reducing the number C of clients participating in every learning round increases the communication efficiency without

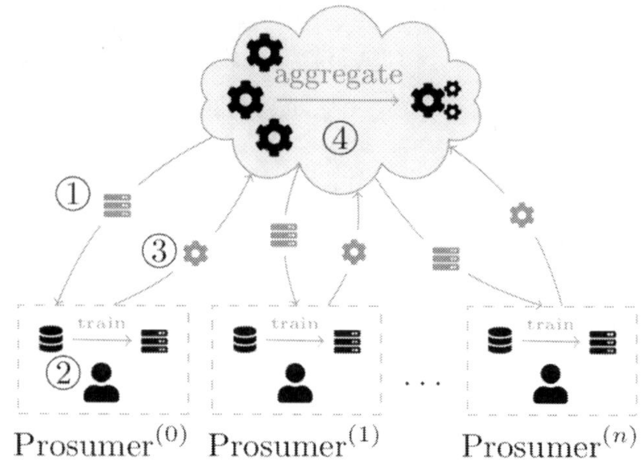

Figure 1. In a Federated Learning approach, all prosumers train their models locally on their own data.

loss of prediction accuracy [8]. So, in every round a prosumer subset $\mathcal{P}'_r \subseteq \mathcal{P}$ with $|\mathcal{P}'_r| = C$ is randomly chosen to take part in the training task in step 2.

Beside the number of prosumers involved in training, the used `aggregate`-function offers additional flexibility. In [8], the author introduces `FedSGD` and `FedAvg`, where the later is the common approach for solving the FL problem by calculating the (weighted) average (often mean) per parameter. Other aggregation approaches are, e.g., federated adaptive optimizers (`FedAdam`, `FedAdagrad`, `FedYogi`) [15], momentum-based variance-reduced technique (`FAFED`) [16], heterogeneity focused (`FedProx` [17], `SCAFFOLD` [18]). There are plenty more proposed `aggregate`-methods, and the related questions in terms of, e.g., applicability, optimality, generalization, are major research topics.

At this point, it is worth noting that additional security mechanism are needed to guarantee some desired security level. Although, FL offers a framework for data-privacy in distributed learning, data leakage or reconstruction attacks are still possible [19]. Privacy enhancing techniques applicable for FL settings are, e.g., differential privacy and homomorphic encryption [20].

In the next section, we give an overview of existing FL research with focus on STLF.

C. Related Work

After describing the FL approach in general, we give an overview of existing FL research conducted in the field of residential STLF. Here, we limit the related work explicitly to (i) residential households and (ii) maximum 24-hour forecast horizon.

A comparison between `FedAvg` and `FedSGD` with different forecast horizons (1 h and 24 h) is given in [21]. They showed that their proposed FL model with `FedAvg` reaches higher accuracy than a centralized and a personalized model.

In [22], the authors compare the forecasting accuracy of a FL model on prosumers involved in training and on hold-out prosumers. They choose this approach to evaluate how well the global model fit for non-participating prosumers. Here, the non-participant prosumers fine tune the pre-trained model for 5 epochs locally. They conclude that this fine tuning step improves the forecast accuracy compared to the global model.

In terms of unbalanced client data distribution, Liu *et al.* proposed the closest approach [23]. Here, clients are divided into 5 groups based on the resolution of their available consumption data ranging from 300 s to 1.800 s.

A hybrid CNN-LSTM model is used in a FL setting in [24]. To handle the consumption heterogeneity, the authors propose a model fine-tuning step after the weight aggregation based on multiple kernel variant of maximum mean discrepancies. Furthermore, all clients are involved in every training and the number of data samples are equal over all clients.

The authors in [25] compare the accuracy of a centralized model with a FL one, a FL plus clustering, and FL plus clustering and subsequently local fine tuning. Here, the last approach reaches the highest accuracy. But, to manage all experiment permutations the evaluations are done with fixed $C = 0.1$.

All of the mentioned related work are summarized with their respective training and model parameters in Table I. It can be seen that the related work in terms of unbalanced data sets is non existing – as far as we know – for the STLF problem on residential prosumer level.

III. Methodology

To evaluate our proposed FL approach, different experiments are conducted in this work. Therefore, we build a federation composed of prosumers represented by household data taken from public available real-world electricity records (see Section III-A).

A. Used Data Set

In this work, residential household data are taken from the `SmartMetersInLondon` [26] data set, which is a refactored version of the "Low Carbon London Project" data. This data set contains electricity consumption records for 5,567 London households between November 2011 and February 2014. In the following, the conducted data preprocessing and preparation steps as well as the selection of suitable households is described.

a) Household Selection: Since the date range differs between prosumers in the data set, only houses with the most overlap are selected. Furthermore, households with more than three consecutive hours of missing values are removed otherwise linear interpolated. In total, 20 households are selected suitable for further usage. The hourly mean electricity consumption is depicted for every day in the training set in Figure 2. Subsequently, the respective consumption data is preprocessed for every selected household in the following.

b) Data Preprocessing: Since the date ranges of available data varies tremendously across all prosumers, we select the time between 1st January 2013 and 28th February 2014 with the most overlapping data. This interval is further divided into train and test data ($\mathcal{D}_{\text{train}}$ and $\mathcal{D}_{\text{test}}$), whereas the whole year 2013 is used for training and the remaining data for testing. This leads to $|\mathcal{D}^{(p)}_{\text{train}}| = 8,760$ and $|\mathcal{D}^{(p)}_{\text{test}}| = 1,416$ samples for every prosumer. For every prosumer, both data sets are normalized individually by $x' = \frac{x-\sigma}{\mu}$, where x' is the transformed consumption time series.

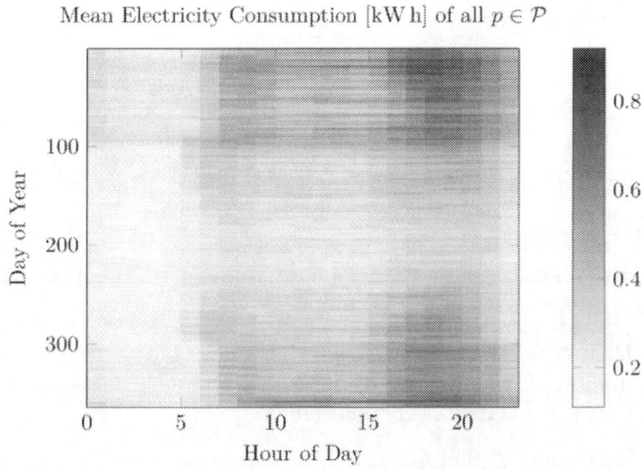

Figure 2. Mean electricity consumption of all selected households from the `SmartMeterInLondon` data set.

c) Look-back and Forecast Horizon: The accuracy of time series forecasting depends on both, the chosen look-back window as well as the forecast horizon. In the related work (Section II-C), those parameter are also chosen variously. Here, our proposed forecasting model uses the last 24 h as input to predict the next 24 h. Although, additional features, e.g., weather, holiday, weekday/weekend, can reduce the forecast error, we restrict our model to the raw consumption values. In [27], we evaluate the FL model with further feature engineering.

After the household selection and necessary preprocessing steps, the used ML model architecture, as well as further detail on the overall development process is given in the next part.

B. System Setting

In this section, we give all the relevant information about the model architecture and used hyperparameters. Afterwards, a definition for different kinds of learning prosumers within the federation based on the ability to store training data is presented. A description of the used federation, as well as the training procedure is given in the third part.

a) Model and Hyperparameters: In this work, we choose a vanilla Multi-Layer Perceptron (MLP) as model architecture, similar to the proposed model in [8]. This architecture allows an easy implementation and training on lightweight devices with limited computational resources. This fully connected MLP has two hidden layers with 200 neurons each and uses a Rectified Linear Unit as activation function.

b) Weak and Strong Prosumer: We introduce the terms *strong* and *weak* prosumer, to describe two different types of

TABLE I. OVERVIEW AND SUMMARY OF RELATED WORK FOR FEDERATED LEARNING (FL) APPROACHES FOR RESIDENTIAL SHORT-TERM LOAD FORECASTING (STLF).

Related Work	#Clients	C	ML-Model	Data Set	Balanced Data	Aggregation
Taïk and Cherkaoui [22]	200	5, 10	LSTM	AUSTIN	yes	FedAVG
Fekri *et al.* [21]	19	6	LSTM	non-public	yes	FedSDG, FedAVG
Liu *et al.* [23]	50	10	iQGRU	AUSTIN	semi	FedAVG
Shi and Xu [24]	10	10	CNN-LSTM	LONDON	yes	FedAVG
Briggs *et al.* [25]	100	0.1	LSTM	LONDON	yes	FedAVG

prosumers based on the amount of available training data. The two types are defined the following way:

Definition 1. Let $p \in \mathcal{P}$ be a prosumer only able to store training data between two consecutive communication rounds, then it is called a *weak* prosumer p_{weak}.

Definition 2. Let $p \in \mathcal{P}$ be a prosumer with no storage limitations, then it is called a *strong* prosumer p_{strong}.

Based on the Definitions 1 and 2, we define the fraction of strong prosumers within a federation as the so-called *strong-prosumer-fraction*:

Definition 3. Let $|p_{\text{weak}}|, |p_{\text{strong}}|$ be the number of weak respective strong prosumers in $\mathcal{P}$, then the strong-prosumer-fraction is defined as $\phi = \frac{|p_{\text{strong}}|}{|p_{\text{weak}}|+|p_{\text{strong}}|}$.

This allows a straightforward distinction between prosumers within a federation and introduces another parameter for the overall training procedure.

c) Training Procedure: For all conducted experiments, with or without strong and weak prosumers, the respective training procedure takes $r = 100$ communication rounds in total. At $r = 0$ the global model's weights w are randomly initialized. After every round, the global model's weights are updated by a weighted FedAvg aggregation function, s.t. $w_{r+1} \leftarrow \sum_{p \in \mathcal{P}'_r} \frac{n_p}{n} w_r^{(p)}$, where n_p, n is the number of sample per prosumer respective the number of all samples. The local weights $w_r^{(p)}$ are calculated locally for every $p \in \mathcal{P}'_r$ in parallel by $w_r^{(p)} \leftarrow w_r - \eta \nabla_w \mathcal{L}(w_r; \mathbf{x}_i, \mathbf{y}_i)$ for a single epoch with a learning rate of $\eta = 0.001$ and the Mean Squared Error (MSE) as loss function $\mathcal{L}(\cdot)$.

To evaluate the proposed FL approach and also to analyze the impact of unbalanced data sets, various experiments are conducted, which are further detailed in the following section.

C. Experiment Settings

The proposed FL approach for residential STLF is evaluated in different experiments. The evaluation is based on the MSE error metric given as $\text{MSE} = \frac{1}{n}\sum_{i=1}^{n}(\hat{y}_i - y_i)^2$, where n is the number of test set samples and $\hat{y}_i, y_i$ is the predicted respective actual consumption value. We run the following three experiments:

I **Benchmark** A centralized model – as well as one local model for every prosumer – is trained over r epochs.

II **Number of Learners** Since a new subset of learning prosumers is selected in every round (see Section II-B), we evaluate the model's forecast accuracy for different number of learners $C = \{1, 3, 5, 7, 10, 20\}$.

III **Strong Prosumer Fraction** With the introduction of weak and strong prosumers, we evaluate our FL approach based on unbalanced data sets. For $C = \{1, 10, 20\}$ the strong-prosumer-fraction $\phi = \{0.05, 0.25, 0.5, 0.75, 1\}$ is considered. Here, the unbalanced data set evolves over the communication rounds $r = \{1, 2, \ldots, 100\}$ by:

$$\text{weak:} \quad \mathcal{D}_r^{(p)} = \mathcal{D}_{r-1:r}^{(p)} \tag{2}$$

$$\text{strong:} \quad \mathcal{D}_r^{(p)} = \mathcal{D}_{0:r}^{(p)}. \tag{3}$$

So, for strong prosumer the training samples increase by $n = \lfloor \frac{|\mathcal{D}|}{r} \rfloor$ in every round, whereas for weak prosumer the samples have a fixed size of n.

The experiments I-III are repeated for $N = 10$ times to handle the randomness via model initialization and prosumer sampling with C, ϕ. Our proposed FL approach is implemented in Python=3.9 with PyTorch and model training was

TABLE II. TEST SET ERROR FOR EXPERIMENT I. MSE IS CALCULATED OVER ALL 20 PROSUMERS

Model	↓ MSE ($\mu \pm \sigma$)	min	max	won
centralized	0.181 ± 0.13	0.030	0.545	3 out of 20
personalized	0.166 ± 0.13	0.021	0.514	17 out of 20

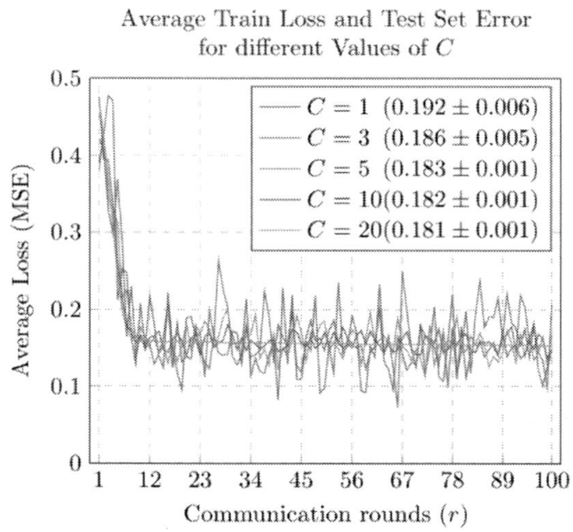

Figure 3. Experiment II: Train loss and test set error with mean and standard deviation over 10 repetitions for different values of C.

executed on a local machine with a Nvidia Geforce RTX 2080 graphic card. The experiments' results are listed in the next section.

IV. Experiment Results & Discussion

The results of the various experiments are presented in the same order as defined in Section III-C. The respective results are provided below, followed by a detailed analysis and discussion.

Figure 3 illustrates the training loss across all communication rounds r as well as the test set error in the legend. For the different values of $C = \{1, 3, 5, 10, 20\}$, the test set error is given as mean with standard deviation over all 10 repetitions. Similar to experiment I, the MSE is calculated over all prosumers $p \in \mathcal{P}$ without individual examination.

For experiment III, results are given in two ways. First, the average training loss over all runs is depicted in Figure 4. Second, Table III list the test set errors. In addition to numerical values over all prosumers, the MSE is also calculated separately for the sets of p_{weak} and p_{strong}. The minimum and maximum MSE values are determined over all 10 runs combined for each combination of C- and ϕ-values.

In this work, a FL approach was proposed for the STLF problem at residential prosumer level. Three experiments were conducted to analyze the impact of unbalanced data distribution among prosumers within the federation.

The first experiment compared a centralized MLP trained on all prosumers' data with a personalized MLP trained individually for each prosumer. Of 20 households in total, 17 times the personalized model reaches a higher accuracy (see Table II). This indicates a strong distribution of consumption behaviour across the prosumers since more data does not guarantee better results.

The second experiment examined the effect of different numbers of learners. As shown in Figure 3, test set errors show minimal variation for $C > 1$, with nearly identical training loss reduction. However, lower C-values introduce more variance, emphasizing trade-off between distribution computational resources and learning efficiency.

In real-world scenarios, training data availability varies among prosumers due to recording and storage capabilities as well as temporal offsets in joining the federation. To address this, the third experiment introduced the distinction between weak and strong prosumers, defined by storage capability. The strong prosumer fraction ϕ represents the proportion of strong prosumers within a federation. Figure 4 indicates slower training convergence with a decreasing number of strong prosumers, irrespective of C-values. However, reducing ϕ to 0.75 or 0.5 did not significantly impact training speed or test set error. This finding is relevant for practical applications, suggesting that not all prosumers need to contribute learning resources to maintain overall performance.

TABLE III. Test set error for experiment III. Error is given as MSE with mean and standard deviation over all 10 repetitions.

C	ϕ	↓ MSE ($\mu \pm \sigma$) all	strong	weak	min	max
1	0.05	*0.215 ± 0.14*	0.192 ± 0.12	0.216 ± 0.14	0.026	0.674
	0.25	**0.193 ± 0.12**	0.209 ± 0.15	0.188 ± 0.11	0.039	0.597
	0.5	0.194 ± 0.12	0.202 ± 0.13	0.186 ± 0.12	0.037	0.565
	0.75	0.196 ± 0.12	0.194 ± 0.12	0.199 ± 0.13	0.038	0.587
	1	0.201 ± 0.13	0.201 ± 0.13	–	0.036	0.626
10	1	*0.223 ± 0.15*	0.142 ± 0.07	0.227 ± 0.15	0.029	0.750
	0.25	0.186 ± 0.12	0.187 ± 0.13	0.186 ± 0.11	0.033	0.540
	0.5	0.184 ± 0.12	0.182 ± 0.12	0.187 ± 0.12	0.038	0.550
	0.75	0.181 ± 0.12	0.185 ± 0.12	0.170 ± 0.10	0.038	0.525
	1	**0.180 ± 0.12**	0.180 ± 0.12	–	0.041	0.527
20	1	*0.198 ± 0.13*	0.205 ± 0.13	0.198 ± 0.13	0.034	0.711
	0.25	0.190 ± 0.12	0.193 ± 0.13	0.189 ± 0.12	0.035	0.591
	0.5	0.183 ± 0.12	0.173 ± 0.11	0.192 ± 0.13	0.040	0.546
	0.75	0.181 ± 0.12	0.172 ± 0.11	0.208 ± 0.12	0.038	0.523
	1	**0.179 ± 0.11**	0.179 ± 0.11	–	0.042	0.516

Note: lowest error is in **bold**, highest in *italic*.

V. Conclusion & Future Work

This work developed a ML-based model for the STLF problem at residential prosumer level. Given that high-resolution electricity consumption data contain behavioral information, data privacy concerns arise when transferring and processing such data. To address this, FL was incorporated as a viable approach to train ML models on distributed data without requiring direct data exchange. Three experiments were designed and conducted to evaluate the proposed FL approach. The results demonstrated that FL can achieve competitive forecasting accuracy while preserving data privacy. The trade-off between the number of learners and computational efficiency was also analyzed, along with the effects of strong and weak prosumers on training convergence and performance.

In future work, we will focus on extending and improving the proposed FL approach. This study primarily addressed unbalanced data sets within a federation, adopting constraints such as a lightweight MLP architecture, state-of-the-art `FedAvg` weight aggregation, and the exclusion of external features. To enhance overall forecasting accuracy, these constraints should be revisited. Preliminary results indicate the utilizing more complex Long-Short Term Memory Neural Network (LSTM) models and incorporating weather information can reduce forecasting errors. Additionally, this study did not explicitly implement a security layer. Future research will explore methods to ensure data privacy and prevent information leakage while integrating insights from this study. Furthermore, the potential of Transformer-based models for STLF remains an unexplored area, warranting future investigation.

Acknowledgment

The research in this work is supported for Alexander Wallis by the Federal Ministry of Education and Research BMBF (FKZ 03FHP212) and for Sascha Hauke by Bayern Innovativ.

References

[1] A. Jain, H. Patel, L. Nagalapatti, N. Gupta, S. Mehta, S. Guttula, *et al.*, "Overview and importance of data quality for machine learning tasks," in *Proceedings of the 26th ACM SIGKDD international conference on knowledge discovery & data mining*, 2020, pp. 3561–3562.

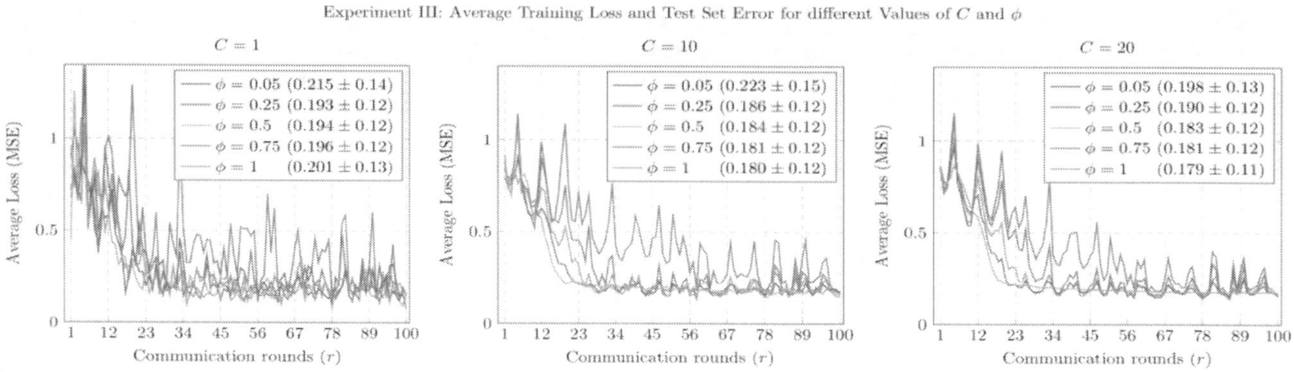

Figure 4. The training loss and test set error for different fractions of strong prosumer ϕ evaluated for $C = 1$ (left), $C = 10$ (middle), and $C = 20$ (right).

[2] H. Habbak, M. Mahmoud, K. Metwally, M. M. Fouda, and M. I. Ibrahem, "Load forecasting techniques and their applications in smart grids," *Energies*, vol. 16, no. 3, p. 1480, 2023.

[3] GDPR, "General data protection regulation," *Regulation (EU) 2016/679 of the European Parliament and of the Council of 27 April 2016 on the protection of natural persons with regard to the processing of personal data and on the free movement of such data, and repealing Directive 95/46/EC*, 2016.

[4] P. L. Ambassa, A. V. Kayem, S. D. Wolthusen, and C. Meinel, "Inferring private user behaviour based on information leakage," *Smart Micro-Grid Systems Security and Privacy*, pp. 145–159, 2018.

[5] G. Wood and M. Newborough, "Dynamic energy-consumption indicators for domestic appliances: Environment, behaviour and design," *Energy & buildings*, vol. 35, pp. 821–841, 2003.

[6] A. Molina-Markham, P. Shenoy, K. Fu, E. Cecchet, and D. Irwin, "Private memoirs of a smart meter," in *Proceedings of the 2nd ACM workshop on embedded sensing systems for energy-efficiency in building*, 2010, pp. 61–66.

[7] J. Chen, H. Yan, Z. Liu, M. Zhang, H. Xiong, and S. Yu, "When federated learning meets privacy-preserving computation," *ACM Comput. Surv.*, vol. 56, no. 12, Oct. 2024, ISSN: 0360-0300. DOI: 10.1145/3679013.

[8] B. McMahan, E. Moore, D. Ramage, S. Hampson, and B. A. y. Arcas, "Communication-Efficient Learning of Deep Networks from Decentralized Data," in *Proceedings of the 20th International Conference on Artificial Intelligence and Statistics*, ser. Proceedings of Machine Learning Research, vol. 54, PMLR, Apr. 2017, pp. 1273–1282.

[9] L. Li, Y. Fan, M. Tse, and K.-Y. Lin, "A review of applications in federated learning," *Computers & Industrial Engineering*, vol. 149, p. 106 854, 2020.

[10] P. M. Mammen, "Federated learning: Opportunities and challenges," *arXiv preprint arXiv:2101.05428*, 2021.

[11] J. Wen, Z. Zhang, Y. Lan, Z. Cui, J. Cai, and W. Zhang, "A survey on federated learning: Challenges and applications," *International Journal of Machine Learning and Cybernetics*, vol. 14, no. 2, pp. 513–535, 2023.

[12] J. Jithish, B. Alangot, N. Mahalingam, and K. S. Yeo, "Distributed anomaly detection in smart grids: A federated learning-based approach," *IEEE Access*, vol. 11, pp. 7157–7179, 2023.

[13] H. Liu, X. Zhang, X. Shen, and H. Sun, "A federated learning framework for smart grids: Securing power traces in collaborative learning," *arXiv preprint arXiv:2103.11870*, 2021.

[14] X. Cheng, C. Li, and X. Liu, "A review of federated learning in energy systems," *2022 IEEE/IAS industrial and commercial power system Asia (I&CPS Asia)*, pp. 2089–2095, 2022.

[15] S. Reddi, Z. Charles, M. Zaheer, Z. Garrett, K. Rush, J. Konečný, *et al.*, *Adaptive federated optimization*, 2021. arXiv: 2003.00295 [cs.LG].

[16] X. Wu, F. Huang, Z. Hu, and H. Huang, *Faster adaptive federated learning*, 2023. arXiv: 2212.00974 [cs.LG].

[17] T. Li, A. K. Sahu, M. Zaheer, M. Sanjabi, A. Talwalkar, and V. Smith, "Federated optimization in heterogeneous networks," *Proceedings of Machine learning and systems*, vol. 2, pp. 429–450, 2020.

[18] S. P. Karimireddy, S. Kale, M. Mohri, S. Reddi, S. Stich, and A. T. Suresh, "Scaffold: Stochastic controlled averaging for federated learning," in *International conference on machine learning*, PMLR, 2020, pp. 5132–5143.

[19] C. Ma, J. Li, M. Ding, H. H. Yang, F. Shu, T. Q. Quek, *et al.*, "On safeguarding privacy and security in the framework of federated learning," *IEEE network*, vol. 34, no. 4, pp. 242–248, 2020.

[20] V. Mothukuri, R. M. Parizi, S. Pouriyeh, Y. Huang, A. Dehghantanha, and G. Srivastava, "A survey on security and privacy of federated learning," *Future Generation Computer Systems*, vol. 115, pp. 619–640, 2021.

[21] M. N. Fekri, K. Grolinger, and S. Mir, "Distributed load forecasting using smart meter data: Federated learning with recurrent neural networks," *International Journal of Electrical Power & Energy Systems*, vol. 137, p. 107 669, 2022. DOI: https://doi.org/10.1016/j.ijepes.2021.107669.

[22] A. Taïk and S. Cherkaoui, "Electrical load forecasting using edge computing and federated learning," in *ICC 2020-2020 IEEE international conference on communications (ICC)*, IEEE, 2020, pp. 1–6.

[23] Y. Liu, Z. Dong, B. Liu, Y. Xu, and Z. Ding, "Fedforecast: A federated learning framework for short-term probabilistic individual load forecasting in smart grid," *International Journal of Electrical Power & Energy Systems*, vol. 152, p. 109 172, 2023.

[24] Y. Shi and X. Xu, "Deep federated adaptation: An adaptative residential load forecasting approach with federated learning," *Sensors*, vol. 22, no. 9, p. 3264, 2022.

[25] C. Briggs, Z. Fan, and P. Andras, "Federated learning for short-term residential load forecasting," *IEEE Open Access Journal of Power and Energy*, vol. 9, pp. 573–583, 2022.

[26] *Smart meters in london*, https://www.kaggle.com/datasets/jeanmidev/smart-meters-in-london, Accessed: 2010-09-30.

[27] A. Wallis, U. Ludolfinger, S. Hauke, and M. Martens, "Towards federated short-term load forecasting," *Internationale Energiewirtschaftstagung (IEWT 2021)*, pp. 1–9, 2021.

Offline Reinforcement Learning Agents for Adaptive Reactive Power Control with Renewable Energy Sources

Tejashri Bhatt, Stephan Balduin, and Eric MSP Veith
Institute for Information Technology
OFFIS e.V.
Oldenburg, Germany
e-mail: {tejashri.bhatt | stephan.balduin | eric.veith}@offis.de

***Abstract*—Conventional reactive power control is typically performed by operators through coordinated switching of power electronic devices. This task is becoming increasingly complex as the integration of renewable energy sources, such as rooftop photovoltaic systems and wind turbines, expands. Maintaining grid stability is critical to ensure energy supply without risking equipment damage. In this context, artificial Reinforcement Learning (RL) agents for reactive power control can assist operators by suggesting actions, though final decisions remain with the operator. High-performing automated RL algorithms are essential for this as they enable execution of complex actions through trial and error, facilitating the adaptable transfer of learning to the real world. While established algorithms, such as Soft Actor-Critic (SAC), Deep Deterministic Policy Gradient (DDPG), Twin-Delayed DDPG and Proximal Policy Optimization (PPO), offer solutions, each has limitations. Training artificial RL agents in real-world power grids is impractical due to the safety-critical concerns, stressing the need for an alternative approach. SAC provides benefits in continuous action space, such as improved exploration and leveraging past experiences, but suffers from long training times. This paper addresses the issue by reducing SAC training periods through the integration of the Behavior Cloning from Observation (BCO) algorithm. This approach enhances performance by initializing SAC with a high-performing, pre-trained Artificial Neural Network (ANN) rather than a random policy, providing a strong starting point while preserving the benefits of SAC.**

Keywords-Smart Grid Management; Reactive Power Control; Artificial Intelligence; Soft Actor-Critic; Behavioral Cloning from Observation; Renewable Energy Integration; Offline Reinforcement Learning.

I. INTRODUCTION

Autonomous systems hold significant potential for power systems, a domain where mismanagement can have extensive societal repercussions. As power systems evolve with increasing dynamic complexity and renewable integration, traditional control methods are becoming inadequate. Autonomous systems provide real-time decision-making, optimize resource allocation and adapt to fluctuating conditions with minimal human input. This transition to autonomy not only enhances operational efficiency but also ensures grid reliability amid rapid technological advancements and energy demands.

Reinforcement Learning (RL), as defined by Sutton et al. [1], is *"learning what to do—how to map situations to actions—so as to maximize a numerical reward signal."* Here, an agent learns by interacting with its environment, receiving feedback in the form of rewards to guide its actions. Two key aspects of RL are its reliance on trial-and-error learning and its consideration of delayed rewards.

Offline RL, or batch RL, trains agents on static datasets, which is advantageous in fields like healthcare, autonomous driving and power systems, where real-time data gathering can be costly or risky [2]. Unlike online RL, which continuously interacts with the environment, offline RL relies on pre-existing datasets to learn policies, minimizing costs and enhancing safety. However, a challenge in offline learning is concerning distributional shift, which arises when the training data differs significantly from real-world scenarios, necessitating high-quality datasets to ensure reliable outcomes.

Behavior Cloning from Observation (BCO) is a notable offline RL algorithm that enables an agent to learn tasks by observations only, without direct access to an expert's actions. By leveraging pre-existing data, BCO mimics expert behavior and iteratively refines policies to perform accurately in complex settings. If the data is optimal, BCO can help overcome common challenges in offline RL, providing robust, efficient learning strategies.

Distributed generators, like Photovoltaic (PV) systems with inverters, offer strategic advantages for reactive power control at key locations. Effective voltage control and Reactive Power Management (RPM) are crucial for power system reliability, typically managed through centralized control by system operators [3]. These operators use comprehensive system data and advanced computer models to make informed decisions, as suppliers do not have direct control over voltage control needs. The operators would benefit from quick assistance in making timely decisions, with a potential Artificial Intelligence (AI) agent suggesting corrective actions while leaving the final decision to the operator. This seamless integration of centralized control and distributed resources could enhance the efficiency and reliability of voltage stability and RPM in power systems.

The primary contributions of this paper are as follows:

1) Addressing controller conflicts arising from physical constraints, such as the impossibility of achieving a 1 voltage magnitude per unit (p.u.) across multiple buses in series simultaneously. In standard Soft Actor-Critic (SAC), an exploration dilemma emerges, as the agent incorrectly assumes actuator independence across buses, where, in reality, actuator behavior is interdependent due to the physical limitations. To address this, we developed a specialized RL agent tailored for this RPM context.

2) Applying BCO to expedite training while maintaining key benefits of off-policy learning and robust exploration. The use of BCO directly aids in overcoming the exploration challenge outlined above, where agents benefit from a structured observational learning approach.
3) Demonstrating the scalability and transferability of the proposed approach for application to more complex tasks within the power systems domain.

This study is constrained by the following factors:

1) The PalaestrAI framework is used for reactive power control implementation.
2) SAC is selected for policy learning due to its suitability for continuous action spaces and superior stability over Deep Deterministic Policy Gradient (DDPG) and Proximal Policy Optimization (PPO) [4].
3) BCO is preferred due to the availability of high-quality MIDAS data (discussed later), making Advantage Weighted Actor-Critic (AWAC) unnecessary [5].
4) Each one-year simulation requires one hour, with additional time for training and testing.

The remainder of this paper is structured as follows: Section II surveys the related work, covering key literature on SAC, BCO and recent advancements in RPM utilizing AI, in addition to open-source tools used in this work such as Midas and palaestrAI. Section III presents the methodology, describing the grid environment and grid code utilized. It also covers the development of three scenarios, experiment setup and the performance metrics applied for evaluation. Section IV presents the results while Section V provides an analysis and discussion of the results for each scenario, as well as ablation experiment. Lastly, Section VI summarizes the main findings and offers potential avenues for future research.

II. Related Work

This section examines the integration of SAC, a state-of-the-art RL method, with BCO, for reactive power control, leveraging historical data and entropy-based learning for enhanced stability. It also reviews advancements in AI-driven reactive power management, identifies research gaps and highlights the open-source tools Midas and palaestrAI for scalable and reproducible simulations.

A. Soft Actor-Critic

The SAC algorithm, introduced by Haarnoja et al. [6], is an off-policy maximum entropy actor-critic framework designed to balance exploration with reward maximization. In entropy-regularized reinforcement learning, entropy—representing randomness in a policy—adds a bonus reward at each step. This motivates agents to explore more by maximizing both the cumulative reward and entropy [7] [8]. This entropy-enhanced approach provides sample-efficient learning, stability and adaptability to complex tasks. For further details on SAC and its implementation, refer to [4] and [7]. The resulting objective function is:

$$\pi^* = \arg\max_{\pi} \mathbb{E}_{\tau\sim\pi}\left[\sum_{t=0}^{\infty} \gamma^t \left(R(s_t, a_t, s_{t+1}) + \alpha H\left(\pi(\cdot|s_t)\right)\right)\right] \tag{1}$$

where $\alpha > 0$ is a regularization factor. The modified value functions $V_\pi(s)$ and $Q_\pi(s,a)$ now include entropy terms, with the following transformations:

$$V_\pi(s) = \mathbb{E}_{a\sim\pi}[Q_\pi(s,a)] + \alpha H\left(\pi(\cdot|s)\right) \tag{2}$$

$$Q_\pi(s,a) = \mathbb{E}_{s'\sim P, a'\sim\pi}[R(s,a,s') + \gamma(V_\pi(s') + \alpha H(\pi(\cdot|s')))] \tag{3}$$

This entropy-enhanced approach provides sample-efficient learning, stability and adaptability to complex tasks. For further details on SAC and its implementation, refer to [4] and [7].

B. Behavior Cloning from Observation

As outlined in the introduction, BCO is a learning approach that enables an agent to mimic expert actions by observing state transitions, bypassing the need for explicit action data and allowing skill acquisition without complete state-action mappings.

The BCO framework begins by initializing policy and model training with an offline dataset, eliminating the need for real-time environment interaction during this stage. Once an initial policy is derived, it is refined through online RL. A common challenge in BCO is the risk of distributional shift, where changes in the offline dataset adversely impact performance, as noted by Prudencio et al. [2]. Thus, a reliable dataset, ideally from expert sources, is crucial for initial training.

In this study, BCO is combined with SAC, which uses entropy-based learning, enhancing the agent's capacity for complex decision-making tasks. Together, BCO and SAC principles contribute to accelerated learning and increased adaptability in reinforcement learning settings. For a more detailed BCO understanding, refer to [9].

C. Progress in Reactive Power Management via AI

Prior research has focused on using artificial intelligence with renewable energy sources to enhance reactive power control and voltage stability in smart grids. For instance, Chandrasekaran et al. [10] propose a hybrid model using solar and wind energy with Artificial Neural Network (ANN) and Distribution Static Synchronous Compensator (DSTATCOM), achieving a voltage profile accuracy of 98.45 % and reducing real power loss by 15 %. Similarly, Utama et al. [11] leverage ANN-based controllers to manage reactive power in the CIGRE Medium Voltage (MV) grid with PV systems, addressing issues such as voltage fluctuations and line congestion in both centralized and decentralized frameworks. To improve smart grid operations, Fiorotti et al. [12] apply a Genetic Algorithm for optimal active and reactive power management, decreasing net present value by 28.15 % and energy costs by 78.16 % by adapting to diverse consumption profiles. However, an approach is needed that can better adapt to complex real-world scenarios and leverage the wealth of historical data effectively. In this context various RL algorithms can be highly advantageous.

Rehman et al. [13] investigate reactive power control in PV inverters through a decentralized actor-critic RL framework, achieving stable voltage control ratios and minimized power loss, with voltage levels consistently within 0.95-1.05 p.u. The authors suggest exploring alternative algorithms for enhanced control in future work. Wolgast et al. [14] employed RL agents for voltage control, with a focus on the impact of environment definitions on performance; however, their study did not explore the influence of advanced RL algorithms. Addressing these limitations, this paper introduces a more advanced algorithm, SAC, over the actor-critic approach and incorporates an ANN for initialization instead of random policy initialization, thereby also implementing BCO.

D. Research Gap

Applying BCO through SAC in RL benchmark environments has, to date, been explored only in one master's thesis by D'Silva et al. [15]. To our knowledge, the application of BCO with SAC specifically in reactive power control for MV grids remains unexplored. Prior research by Dey et al. [16] has utilized BCO with PPO for building energy control; however, the on-policy nature of PPO limits its efficiency in multi-task settings. By substituting PPO with SAC, an off-policy algorithm, this study overcomes these constraints, enabling more efficient training and enhanced grid management performance. This approach extends BCO to grid integration scenarios, leveraging abundant historical data and an ANN-trained policy to initialize SAC for managing reactive power control in an MV grid.

This research investigates whether constructing BCO by integrating the SAC algorithm with ANN policy initialization can reduce training latency for SAC agents in achieving effective reactive power control.

E. Open-Source Tools

1) Midas: Midas is an open-source framework that allows easy configuration of a power system co-simulation [17] and uses the co-simulation framework Mosaik as backend [18]. It features various time series for consumers and producers, weather time series and simulation models of renewable energy sources like photovoltaic, wind and biogas, as well as a battery and a cold warehouse. Midas is configured with scenario files in YAML format, which specify what kind of load or generation is connected to certain buses in the grid. The framework provides a seamless integration into palaestrAI, which is described in the following section.

2) PalaestrAI: The palaestrAI framework is instrumental in simulating real-world scenarios for electricity grid integration, offering a comprehensive set of components. It is designed to implement the Adversarial Resilience Learning (ARL) methodology by Veith et al. [19]. It encompasses various packages, emphasizing a reliable experimentation process through experiment definitions and proper data storage. Users can easily create experiment files to achieve reproducible simulations, defining environments, agents and their parameters for experimentation within palaestrAI.

F. Objective Function

The reward function of the RL agent used for this study is composed of four main components, each associated with specific weighting variables represented by the world state of the system in terms of voltage levels across all buses, buses particularly under control, the bus status based on the impact of grid code violations and the quantity of real power production within the grid relative to total demand. For more details on the formulation of objective function please refer [20].

III. METHODOLOGY

All experiments in this paper utilize the palaestrAI framework from Veith et al. [19] focusing on a single agent. Each experiment includes distinct training and testing phases, with a maximum simulated duration of one year and a 15-minute interval for each step. The overall process flow for methodology is depicted in Figure 2.

A. Grid Environment

In this study, a 20 kV MV grid is connected to a 110 kV transmission network, with a total power capacity of 2000 kW. The grid is modeled after the CIGRE MV benchmark grid [21], comprising 14 buses, each equipped with a PV generator with randomly assigned output for variability. Weather data from Bremen, Germany (see Figure 1) and static load time series from the Midas project simulate realistic conditions [17]. Additionally, commercial loads, such as a supermarket and a small hotel, enrich the grid's complexity.

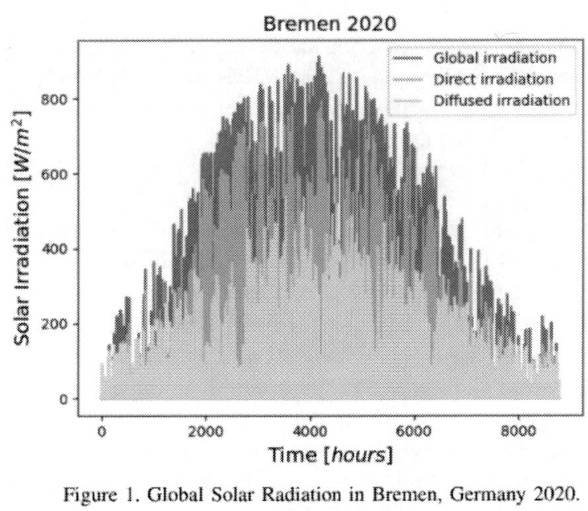

Figure 1. Global Solar Radiation in Bremen, Germany 2020.

This paper follows technical guidelines (VDE-AR-N 4110) published by the German Association for Electrical, Electronic & Information Technologies for MV grids in Germany, ensuring compliance with standards for the safe integration of renewable energy. The reactive power limits of each PV inverter are strictly followed, with any excess set points automatically adjusted to the maximum allowed value, using Volt-VAR control.

The determination of the bus's operational status adheres to the rules outlined in the grid code DIN 50160 for medium voltage grid, also detailed in Table I and adopted from [22].

TABLE I. GRID CONSTRAINTS FOR MEDIUM VOLTAGE GRID.

Grid constraints	Limits
Bus ΔV must be within $\leq$	0.1 p.u./min
Loads to sustain ΔV of $\leq$	0.02 p.u./min
Generators must endure ΔV of $\leq$	0.05 p.u./min
Line load must be $\leq$	100 %

B. Reactive Power Management

1) Building Experiments: The challenge of finding reliable data is mitigated by using the Q-Controller (4) as the expert data source, validated by Ju et al. [23] for stability in reactive power control. The reactive power controller is implemented using the palaestrAI framework, with SAC chosen for its efficiency in continuous action spaces, outperforming DDPG and PPO [4]. BCO is selected for its simplicity. Each simulation spans one year, with data collected in 15-minute intervals, incorporating both a one-year training period and a subsequent one-year testing phase.

The primary objective is to construct an ANN architecture focused on reactive power control, utilizing data generated from the Midas simulation, used as the offline data, to enable BCO in a later stage. Before designing the ANN, the dataset is prepared using (4) to generate q_{t+1} based on the q_t and V_t values. This equation, adapted from Ju et al. [23], provides stability and convergence in reactive power control and serves as the foundation for controlling reactive power set points in the grid.

$$q_{t+1} = [q_t - D(V_t - 1)]^+ \tag{4}$$

Using the simulated data, consisting of voltage (V_t) and reactive power (q_t) values, the effectiveness of (4) is evaluated by generating q_{t+1} values and comparing them to the original q_t values. This analysis spans all 14 buses over a year. The neural network is then developed using V_t[p.u.] and q_t[MVAr] as inputs (x) and q_{t+1}[MVAr] as the output (y).

14 datasets, ranging from 2500 to 34050 data points in increments of 2500, were generated from simulated data. For each dataset, a model is developed (e.g., $model_{2500}$ for 2500 data points, $model_{5000}$ for 5000 data points, etc.) with default hyperparameters initially, repeating this process across all datasets. The dataset is divided into training and testing sets, allowing for an assessment of model performance.

Along with generating 14 models for varying dataset sizes, hyperparameter optimization is performed using a randomized search method from Pedregosa et al. [24], to enhance performance of each model. This optimization targets five out of six hyperparameters—activation function, learning rate, number of neurons per layer, number of layers and batch size—while the number of episodes is fixed at 100. This choice is based on the observed trend in loss versus episode plots for the testing sets, where no significant reduction in loss occurs beyond 100 episodes. The evaluation of the models is based on mean, variance, Root Mean Square Error (RMSE) and Coefficient of Determination (R2). The model with the most favorable metrics—a mean similar to the original data, reduced variance, lower RMSE and higher R2—is selected for further experimentation.

These models are central to two of the three experiments explained later in Section III-B2, highlighting the importance of establishing a resilient network at this stage. Following this, three experimental approaches are explored: Supervised Experiment (SUP), SAC algorithm and a combination of SAC and ANN, thereby implementing BCO as depicted in Figure 2.

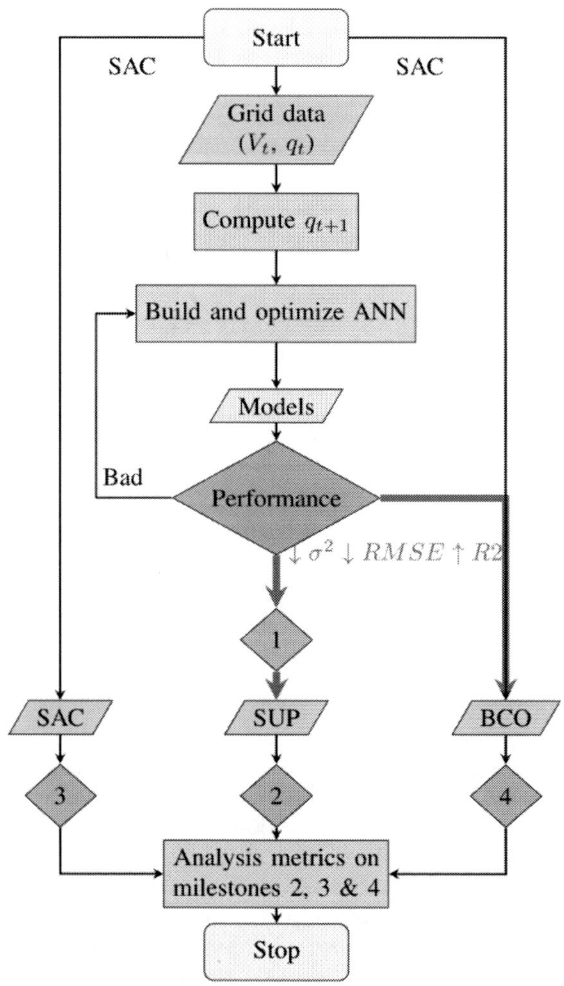

Figure 2. Process Flow Chart for Methodology.

2) Experiment Setup: The structure of the three experiments is presented in Table II. These experiments are designed to build progressively upon each other.

The first experiment, referred to as the SUP experiment, is the simplest and is based entirely on expert knowledge derived from (4), without any RL training process. This expert knowledge is transferred to the ANN using supervised learning. At each time step, sensor values (V_t, q_t) are input into ANN, which produces an output used to set the value of q_{t+1}. Therefore, in this setup, the ANN acts as an actuator without policy initialization, interacting with a medium-voltage (MV) grid environment.

In the second experiment, the SAC algorithm, a well-established method in RL, serves as a baseline for comparison. SAC encourages exploration in RL and operates as an off-policy algorithm, leveraging prior knowledge. However, due to random policy initialization, the SAC algorithm can experience prolonged simulation times in the learning phase.

The third experiment addresses the limitations of both earlier approaches by combining the SAC algorithm with the ANN constructed from the expert knowledge, as the initial policy instead of a random initialization. The objective of this experiment is to integrate the characteristics of the two previous experiments.

This approach, known as BCO, mitigates the initial slow learning problem observed in the SAC experiment, as the ANN provides a more effective starting point. Consequently, the third experiment is expected to outperform both the SUP experiment, which lacks a RL simulation process and the SAC experiment, which suffers from random policy initialization, as illustrated in Table II.

The fourth experiment is an ablation study that compares the voltage profiles and reward values across all three previously mentioned experiments, considering both single bus and multiple bus scenarios.

TABLE II. DESCRIPTION OF AGENT CONFIGURATIONS FOR EACH EXPERIMENT ON THE MV GRID ENVIRONMENT.

Experiment	SUP	SAC	BCO
Objective	Reward calculation		
Sensor	V_t, q_t, %load & in-service status		
Actuator	q_{t+1}		
Policy Initialization	None	Random	ANN

3) Performance Metrics:

a) Models

In the SUP and BCO experiments, a total of 14 models, each trained on datasets of varying sizes, are analyzed to identify the model that achieves optimal performance with minimal data usage. Key evaluation metrics for the ANN models include mean, variance, R2 and RMSE. These metrics will guide the selection of sample-efficient models for both SUP and BCO experiments. In the SUP experiment, the models are used as actuators and in BCO, the models are used in policy initialization as mentioned in Section III-B2, experiment setup.

b) Experiments

The evaluation of experiments is based on six key metrics: voltage stability, adherence to voltage limits, high-reward performance, consistency, sample efficiency and robustness under controlling multi-bus scenarios.

(i) Both voltage and reward values are tracked over time to assess stability, compliance and overall performance.

(ii) Performance Consistency is ensured by repeating each experiment four times and comparing the outcomes across trials, focusing on voltage and reward stability.

(iii) Sample Efficiency is evaluated differently for SUP and SAC. In the SUP experiment, it is identifying the model that achieves the best performance with the least data. The model providing the reactive power set point is trained on various datasets of differing sizes (from 2500 until 35040 data points), with each training repeated four times to evaluate performance consistency. The selection of dataset sizes was intended to examine the impact of sample size on model performance and efficiency. In SAC experiments, sample efficiency is calculated using the following two criteria:

- The rate of change of reward:

$$\eta_s = \frac{dR}{dt} \tag{5}$$

- The area under the reward versus time plot.

(iv) Robustness is evaluated by simultaneously managing two buses, while other buses maintain a $\cos\phi = 0.9$ so the reactive power depends on the PV power injection, which significantly impacts grid dynamics. A model was developed using 35040 data points, incorporating four inputs—voltage and reactive power for each bus—and two outputs for the reactive power of both buses. This model enables the simulation of concurrent control of the buses. Scenarios are deemed robust when the voltage remains within the range $0.85 \leq V_{b,t}[\text{p.u.}] < 1.15$ for all buses b and time t, since beyond these limits buses will get disconnected from the grid and when the rewards are within the range of $0.90 \leq R_{b,t}[\text{-}] < 1.00$. The percentage of values within these specified limits is calculated to assess robustness.

(v) Comparison of Experiments uses a rolling average analysis over ten days to compare voltage and reward outcomes, while sample efficiency is assessed through reward rate of change and Area Under the Curve (AUC) values as mentioned earlier.

IV. RESULTS

This section presents the performance evaluation of different reinforcement learning approaches for voltage control in a power grid scenario. The effectiveness of the SUP, SAC and BCO methods is analyzed in maintaining voltage stability and optimizing rewards under both single and multi-bus control scenarios. The evaluation is conducted by measuring the percentage of voltage and reward values that fall within the acceptable ranges of $0.85 \leq V_{b,t}[\text{p.u.}] \leq 1.15$ and $0.90 \leq R_{b,t}[-] \leq 1.00$, respectively, ensuring reliable grid operation. The results of these experiments are summarized in Tables III and IV, where the former illustrates the single bus performance while the latter demonstrates the robustness against controlling two buses.

A. SUP Experiment

The sample efficient model for single bus scenario ensures $99.9\,\%$ of voltage and $97.0\,\%$ of rewards fall within the desired ranges as shown in Table III. Buses 5 and 11 are chosen for evaluation of robustness against controlling two buses simultaneously for all the three scenarios. $99.83\,\%$ of voltage values and $70.80\,\%$ of reward values are within the specified ranges.

B. SAC Experiment

Since SAC uses random initialization and no models, model optimization is not applicable. Therefore, simple SAC training runs are carried out. For single bus scenario, SAC has $99.9\,\%$ of voltage values within $0.85 \leq V_{b,t}[p.u.] \leq 1.15$ and $47.0\,\%$ of reward values within $0.90 \leq R_{b,t}[-] \leq 1.00$, as shown in Table III. The robustness against controlling two buses is demonstrated by $100.00\,\%$ of voltage values and $70.59\,\%$ of reward values falling within their respective desired ranges, as seen in Table IV.

C. BCO Experiment

The sample efficient model for single bus scenario ensures $99.9\,\%$ of voltage and $65.60\,\%$ of rewards fall within the desired ranges (see Table III). For the robustness of BCO experiment involving simultaneous control of two buses, $100\,\%$ of the voltage values and $81.56\,\%$ of the reward values fall within the respectively desirable ranges.

V. Discussion

This section presents key findings from the experiments, focusing on the performance of the proposed approach across different evaluation metrics.

A. SUP Experiment

The analysis based on voltage violations and reward performance concluded that $model_{5000}$ outperformed those trained on larger datasets, demonstrating the trade-off between dataset size and model efficiency. This shows the potential advantages of smaller datasets in achieving improved model performance and robustness against over-fitting.

During the robustness analysis, both buses exhibited significant performance issues, with voltage levels dropping to zero in 59 instances (0.17 %), violating the grid code requirements. Although the overall objective function remained high (ranging from 0.7 to 1) for most of the simulation period, these voltage drops adversely affected the reward, reducing it to approximately 0.5 in certain instances. In total, 10232 occurrences of rewards below 0.9 accounted for $29.20\,\%$ of the total time steps.

These sudden voltage spikes, observed in Figures 4a and 4b, primarily occurred during periods of high solar irradiation from April to September. These spikes are likely due to solar input exceeding the voltage limits set by the grid code. As shown in Figure 4c, the reward follows an opposing pattern to the voltage spikes, with the reward decreasing for high voltage deltas, in accordance with the grid code. Consequently, the current model's robustness is questioned, as it fails to maintain compliance with the voltage limits across multiple buses, despite achieving reward performance for $70.80\,\%$ of the total time steps. For the robustness analysis, a consistency check through multiple simulation runs was not conducted to minimize the overall number of simulations performed.

An overview of the evaluated criteria is provided in Table III. The values displayed are for the best performing models only.

B. SAC Experiment

In the single-bus scenario, the performance of the SAC experiment is the weakest among the three experiments evaluated. While the voltage values largely remain within the desired range, as presented in Table III, only $47.0\,\%$ of the reward values fall within the range $0.90 \leq R_{b,t}[-] \leq 1.00$, indicating poor adherence to the reward function. Considering the robustness of the SAC algorithm in multi-bus scenarios, it demonstrates impressive robustness. The average voltage performance of both buses, consistently maintains values between 0.96 and 1.03, with no grid code violations. Although reward values remain high, only $70.56\,\%$ of the values are between 0.9 and 1. This consistent voltage stability and moderately satisfactory reward performance demonstrate the resilience of the SAC algorithm (see Table IV).

C. BCO Experiment

Performance of $model_{30000}$ is the most sample-efficient performance based on voltage violations and reward values whereas the robustness analysis reveals that both buses exhibit higher voltages during sunny periods, likely due to increased solar irradiation, which affects reward acquisition but maintains grid code compliance. In contrast, voltages during the less sunny months remain within the desired range, with zero grid code violations observed. Overall, this scenario demonstrates a $0.17\,\%$ improvement in performance compared to the initial SUP experiment, suggesting that the ANN model, enhanced by the SAC algorithm, slightly outperforms the standalone SUP experiment.

The voltage distribution is primarily concentrated between 1 and 1.02 p.u., with some outliers between 1.02 and 1.06 p.u.. Interestingly, the reward performance reflects the inverse of the solar irradiation profile, with most values between 0.9 and 1 and $81.56\,\%$ falling within this range (see Table IV). This indicates a $10.97\,\%$ performance improvement for BCO over SAC, as both maintain $100\,\%$ compliance with voltage standards.

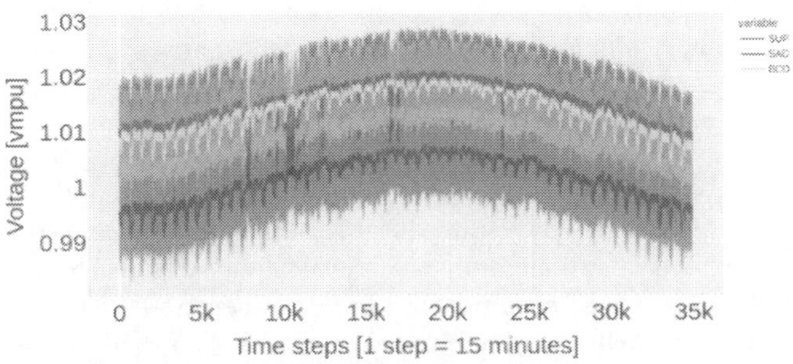

(a) Voltages of Single Bus.

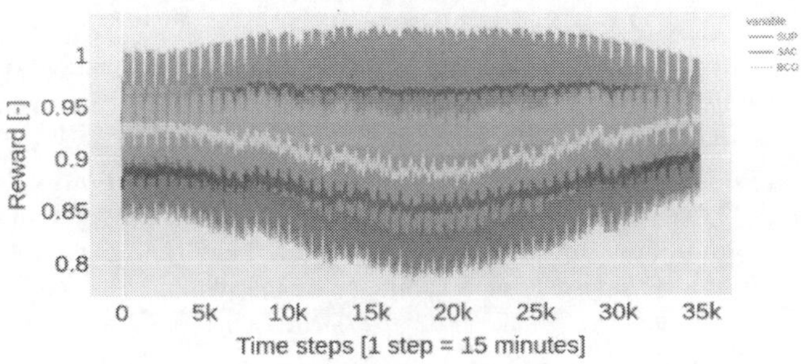

(b) Rewards for Single Bus.

Figure 3. Comparison of Performances for Single Bus.

TABLE III. PERFORMANCE EVALUATION FOR SINGLE BUS

Metrics	SUP	SAC	BCO
Voltage [%]	99.9	99.9	99.9
Reward [%]	97.0	47.0	65.6
Sample efficiency			
Data points	5000	N/A	30000
SAC algorithm	N/A		
Slope	N/A	-0.0020	-0.0024
AUC	N/A	22.45	22.75

TABLE IV. ROBUSTNESS EVALUATION FOR TWO BUSES

Metrics	ANN	SAC	BCO
Voltage [%]	99.83	100.00	100.00
Reward [%]	70.80	70.59	81.56

D. Ablation Experiment

Table III represents a summary of the performance of single bus scenario for all the three experiments. The voltage occurrences within $0.85 \leq V_{b,t}[p.u.] \leq 1.15$ and reward occurrences within $0.90 \leq R_{b,t}[-] \leq 1.00$ are reported as percentages. Sample efficiency is measured by data points required for high model performance and by reward rate of change over the first 25 training hours for the SAC algorithm. Only the best-performing models' values are shown in this table.

1) Rolling Average for Single and Multiple Bus Cases: Figures 3a and 3b present the rolling voltage and reward averages respectively over the 34050 time steps for each experiment in the single bus case. The SUP experiment is constructed using 5000 data points model, while BCO experiment uses model generated from 30000 data points. In the multiple-bus case, Figures 4a and 4b show the voltage performances for Buses 5 and 11, respectively and Figure 4c demonstrates the reward performance. The rolling average approach smooths out outliers, enhancing the visibility of performance trends across time.

Figure 3a shows that all three experiments, SUP, SAC and BCO, maintain stable voltage performance. However, Figure 3b reveals that both SUP and BCO outperform SAC in reward collection, benefiting from the ANN model and initialization advantage, respectively. This model initialization, along with the high entropy inherited from SAC algorithm, consistently keeps BCO experiment ahead in reward collection compared to the SAC experiment. It is intriguing to observe that the performance of SUP experiment remains consistently higher than the other two experiments, making it the best performing experiment in the single bus case.

Figures 4a and 4b depict the voltage performances for Buses 5 and 11 in the multi-bus scenario. SUP shows frequent voltage drops, reflecting poor voltage management across both buses. However, both BCO and SAC experiments exhibit consistent voltage control, without any grid code violations. Short periods of voltage exceeding 1.02 p.u. reduce the rewards for SAC and BCO, but their overall performance remains excellent with performance improving with time. Figure 4c illustrates reward behavior for the multi-bus case, where BCO shows the best overall performance. SAC exhibits more resilient behavior than BCO during the high solar irradiation period, maintaining slightly higher rewards. BCO outperforms SAC in collecting more rewards during the lower solar irradiation period. Despite occasional voltage irregularities, BCO's performance highlights the effectiveness of using the ANN model for initialization, which consistently gives it a head start in reward collection, compared to SAC. Overall, BCO outperforms SAC by collecting $10.97\,\%$ higher rewards. SUP's performance, while strong in the single-bus case, falls short in multi-bus control scenarios.

2) Sample Efficiency Comparison: A comparison between SUP and BCO highlights the differences in sample efficiency in Table III. For the comparison of SAC and BCO sample efficiency, Figure 5 and Table V are used, as described in Section III, Methodology.

Referring to Table III, SUP, using only 5000 data points versus BCO's 30000, is more sample efficient, likely due to SAC's enhanced exploration capability. The complexity of BCO's combined model requires a larger dataset to capture patterns.

Considering (5), BCO shows a $20\,\%$ steeper slope and a $1.34\,\%$ larger AUC than SAC during the initial training phase (Figure 5), confirming its superior sample efficiency, summarized in Table V.

3) Consistency and Robustness Assessment: Consistency is evaluated by carrying out 4 repetitions for each experiment. Figures 3 and 4 show an average of these four repetitions.

TABLE V. COMPARISON OF SLOPE AND AUC IN THE FIRST 25 HOURS OF SAC AND BCO TRAINING.

Criteria	SAC	BCO	% Difference
Slope	-0.0020	-0.0024	20.00
AUC	22.45	22.75	1.34

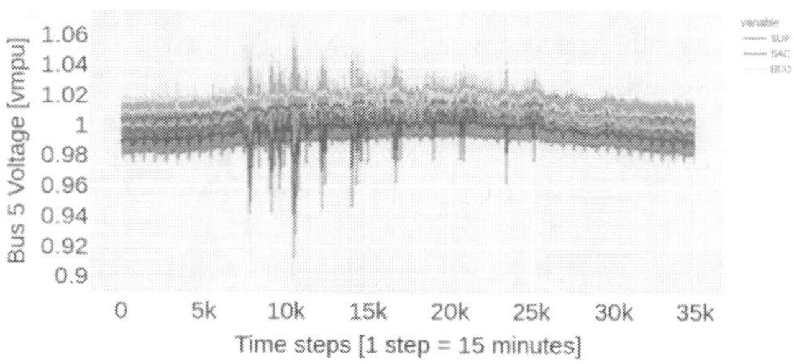

(a) Voltages of Bus 5.

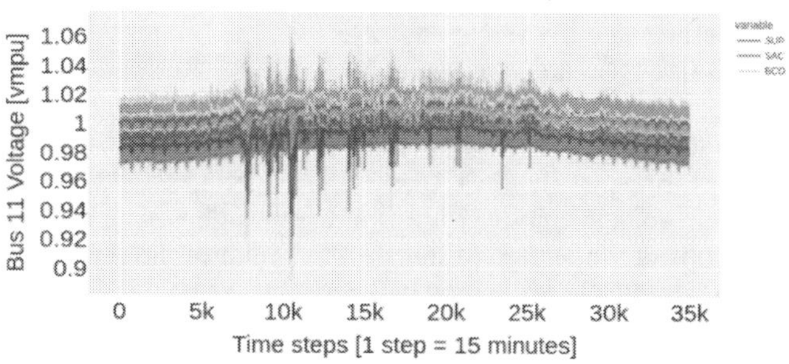

(b) Voltages of Bus 11.

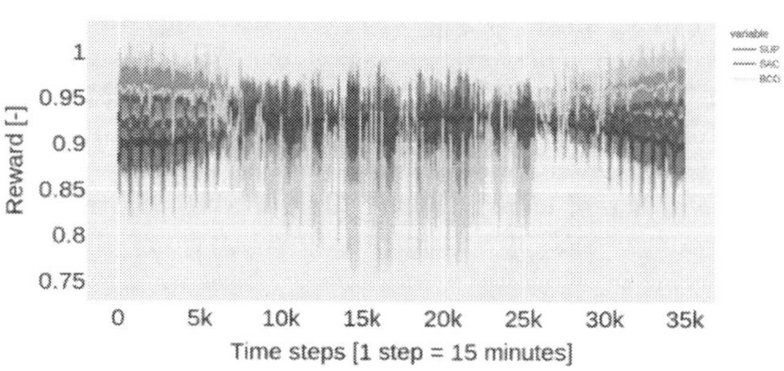

(c) Rewards for Two Buses.

Figure 4. Comparison of Performances for Two Buses.

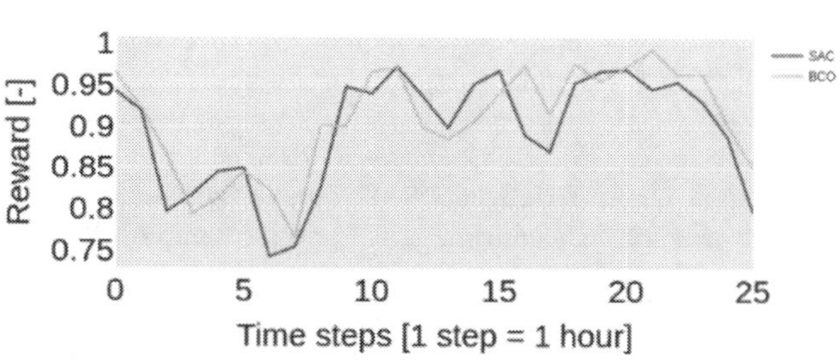

Figure 5. Sample Efficiency: Comparison of reward collection for initial 25 hours of training phase for SAC and BCO experiments.

Robustness against controlling two buses is evaluated based on the criteria mentioned in Section III, Methodology .

All three experiments demonstrate a certain level of consistency, although a minor inconsistency arises due to the distributional shift caused by employing different seed values for each of the four repetitions.

In terms of robustness in managing multiple buses, assessed by the percentage of occurrences where voltage and reward values stay within specified limits, BCO demonstrates the highest resilience. It is followed by SAC, which shows $10.97\,\%$ lower reward collection and then SUP, which, although gradually becoming more robust over time, exhibits lowest stability in multi-bus control due to frequent voltage violations and extended learning times.

VI. Conclusions and Future Work

In the single-bus control scenario, the SUP experiment exhibits high sample efficiency for model by effectively utilizing a smaller dataset, achieving strong reward collection and stable voltage control. However, in multi-bus scenarios, it struggles with voltage stability and reward collection, bringing attention to limitations in handling more complex environments. Table VI can be referred for a summary of the results.

TABLE VI. OVERVIEW OF THE RESULTS.

	Evaluation Criteria	SUP	SAC	BCO
Single Bus	Voltage Stability	✓	✓	✓
	Reward Collection	✓	×	-
	Sample Efficiency:			
	Model	✓		×
	SAC algorithm		×	✓
Two Buses	Voltage Stability	×	✓	✓
	Reward gained	-	×	✓

The SAC experiment demonstrates reliable voltage stability in both single- and multi-bus experiments but falls short in reward collection compared to the other methods, due to its weaker model initialization. Despite this, SAC's algorithm proves robust in managing complex grid conditions, although at the cost of sample efficiency.

The BCO experiment emerges as the best-performing method overall. It maintains superior voltage stability across both scenarios, shows excellent sample efficiency for the SAC algorithm and achieves higher reward collection, making it the most effective solution for both simple and complex grid control tasks.

Future research could enhance performance by exploring advanced neural network architectures and alternative hyperparameter optimization methods. Additionally, modifying the objective function to penalize voltage violations and conducting more repetitions could improve adherence to grid standards and strengthen result confidence.

ACKNOWLEDGMENTS

This work has been funded by the German Federal Ministry of Education and Research (BMBF) under grant AGenC (01IS22047C). The work on the ARL agent's architecture is funded by the BMBF under grant no. 01IS22071.

REFERENCES

[1] R. S. Sutton and A. G. Barto, *Reinforcement learning: An introduction*. MIT press, 2018.

[2] R. F. Prudencio, M. R. Maximo, and E. L. Colombini, "A survey on offline reinforcement learning: Taxonomy, review, and open problems", *IEEE Transactions on Neural Networks and Learning Systems*, pp. 1–8, 14, 2023.

[3] B. Kirby and E. Hirst, "Ancillary service details: Voltage control", Oak Ridge National Lab.(ORNL), Oak Ridge, TN (United States), Tech. Rep., 1997.

[4] T. Haarnoja, H. Tang, P. Abbeel, and S. Levine, "Reinforcement learning with deep energy-based policies", in *International conference on machine learning*, PMLR, 2017, pp. 1352–1361.

[5] A. Nair, A. Gupta, M. Dalal, and S. Levine, "Awac: Accelerating online reinforcement learning with offline datasets", *arXiv preprint arXiv:2006.09359*, 2020.

[6] T. Haarnoja, A. Zhou, P. Abbeel, and S. Levine, "Soft actor-critic: Off-policy maximum entropy deep reinforcement learning with a stochastic actor", in *International conference on machine learning*, PMLR, 2018, pp. 1861–1870.

[7] OpenAI, *Soft actor-critic (sac)*, https://spinningup.openai.com/en/latest/algorithms/sac.html, Accessed: 2025-02-13, 2023.

[8] Z. Ding and H. Dong, "Challenges of reinforcement learning", *Deep Reinforcement Learning: Fundamentals, Research and Applications*, pp. 249–272, 2020.

[9] F. Torabi, G. Warnell, and P. Stone, "Behavioral cloning from observation", *arXiv preprint arXiv:1805.01954*, 2018.

[10] K. Chandrasekaran, J. Selvaraj, C. R. Amaladoss, and L. Veerapan, "Hybrid renewable energy based smart grid system for reactive power management and voltage profile enhancement using artificial neural network", *Energy Sources, Part A: Recovery, Utilization, and Environmental Effects*, vol. 43, no. 19, pp. 2419–2442, 2021.

[11] C. Utama, C. Meske, J. Schneider, and C. Ulbrich, "Reactive power control in photovoltaic systems through (explainable) artificial intelligence", *Applied Energy*, vol. 328, p. 120 004, 2022.

[12] R. Fiorotti *et al.*, "A novel strategy for simultaneous active/reactive power design and management using artificial intelligence techniques", *Energy Conversion and Management*, vol. 294, p. 117 565, 2023.

[13] A. u. Rehman *et al.*, "Artificial intelligence-based control and coordination of multiple pv inverters for reactive power/voltage control of power distribution networks", *Energies*, vol. 15, no. 17, p. 6297, 2022.

[14] T. Wolgast and A. Nieße, "Learning the optimal power flow: Environment design matters", *Energy and AI*, vol. 17, p. 100 410, 2024, ISSN: 2666-5468. DOI: https://doi.org/10.1016/j.egyai.2024.100410.

[15] A. D'Silva, "Integrating behavioral cloning into a reinforcement learning pipeline", Available at https://www.politesi.polimi.it/handle/10589/208354, M.S. thesis, Politecnico di Milano, 2022.

[16] S. Dey, T. Marzullo, X. Zhang, and G. Henze, "Reinforcement learning building control approach harnessing imitation learning", *Energy and AI*, vol. 14, p. 100 255, 2023.

[17] S. Balduin, E. M. Veith, and S. Lehnhoff, "Midas: An open-source framework for simulation-based analysis of energy systems", in *International Conference on Simulation and Modeling Methodologies, Technologies and Applications*, Springer, 2022, pp. 177–194.

[18] A. Ofenloch *et al.*, "Mosaik 3.0: Combining time-stepped and discrete event simulation", in *2022 Open Source Modelling and Simulation of Energy Systems (OSMSES)*, IEEE, 2022, pp. 1–5.

[19] E. Veith *et al.*, "Palaestrai: A training ground for autonomous agents", in *Proceedings of the 37th annual European Simulation and Modelling Conference. EUROSIS*, 2023.

[20] E. M. Veith and E. Frost, "Cover me: Safeguarding multi-agent system with deep reinforcement learning for resilient grid operation", in *Proceedings of the 38th annual European Simulation and Modelling Conference*, Eurosis, Oct. 2024, pp. 3–4.

[21] T. Force *et al.*, "Benchmark systems for network integration of renewable and distributed energy resources", *no. April*, p. 63, 2014.

[22] E. Veith, T. Logemann, A. Wellßow, and S. Balduin, "Play with me: Towards explaining the benefits of autocurriculum training of learning agents", in *2024 IEEE PES Innovative Smart Grid Technologies Europe (ISGT EUROPE)*, Dubrovnik, Croatia: IEEE, 2024, pp. 1–5. DOI: 10.1109/ISGTEUROPE56780.2023.10408277.

[23] P. Ju and X. Lin, "Adversarial attacks to distributed voltage control in power distribution networks with ders", in *Proceedings of the Ninth International Conference on Future Energy Systems*, 2018, pp. 291–302.

[24] F. Pedregosa *et al.*, "Scikit-learn: Machine learning in Python", *Journal of Machine Learning Research*, vol. 12, pp. 2825–2830, 2011.

A Systems Dynamics Analysis of Communication Technology Integration in Complex Transactive Energy Systems

A. Zarei, E. Honu, M. Mansouri, and P. Odonkor*
Department of Systems and Enterprises
Stevens Institute of Technology
Hoboken, USA
email: azarei1@stevens.edu, ehonu@stevens.edu, mmansour@stevens.edu, podonkor@stevens.edu

***Abstract*—The transition to renewable energy and the increasing integration of Distributed Energy Resources (DERs) have transformed power systems into complex sociotechnical networks. Transactive Energy Systems (TES) offer a decentralized framework to facilitate energy transactions, yet their implementation faces challenges related to communication, cybersecurity, scalability, and environmental trade-offs. This study explores the potential of Long Range (LoRa) communication technology as an enabler of TES, using a systems dynamics approach to analyze its systemic impacts. Through stakeholder analysis and causal loop modeling, the research identifies key reinforcing and balancing feedback mechanisms shaping TES adoption. Findings highlight LoRa's benefits, including enhanced scalability, reduced operational costs, and improved resilience, while also revealing dynamic trade-offs in cybersecurity and network congestion. By applying systems thinking to emerging technologies, this work provides valuable insights for advancing sustainable and resilient energy systems.**

***Index Terms*—Transactive Energy Systems; LoRa Technology; System Dynamics; Sustainable Energy Solutions.**

I. Introduction

The global energy landscape is shifting in response to rising demand, rapid urbanization, and the need to address climate change [1]–[3]. As societies transition to sustainable energy sources, power systems must integrate renewable energy while ensuring grid stability and reliability. This challenge is particularly pronounced in developing regions, where growing energy needs intersect with ambitious sustainability goals [4].

Traditional power systems, designed for one-way electricity flow from centralized generators to consumers, are increasingly inadequate for managing Distributed Energy Resources (DERs), such as solar panels, wind turbines, and energy storage systems. While DERs offer environmental and technical benefits through localized energy production, they also introduce variability and bidirectional power flows that complicate grid management [5].

Transactive Energy Systems (TES) provide a framework for addressing these complexities. Defined by the GridWise Architecture Council as "a set of economic and control mechanisms that allow the dynamic balance of supply and demand across the entire electrical infrastructure using value as a key operational parameter" [6], TES establish decentralized marketplaces where energy transactions are guided by real-time supply, demand, and grid conditions [7] [8]. This approach has shown potential for improving affordability, efficiency, and sustainability in power systems [9].

However, TES rely on robust communication infrastructure to function effectively [10] [11]. Many existing communication technologies struggle to meet TES requirements, particularly in terms of energy efficiency, range, and security. These limitations can hinder deployment, especially in areas with dispersed energy resources or weak infrastructure. Additionally, TES face challenges related to data security, privacy, and system resilience [12] [13].

Low-Power Wide-Area Network (LPWAN) technologies, particularly LoRa (Long Range), offer a promising solution to these communication challenges [14] [15]. LoRa enables long-range, low-power communication, making it well-suited for distributed energy systems [16] [17]. Understanding how LoRa's characteristics align with TES requirements is critical for advancing sustainable energy systems [18].

This paper examines the relationship between LoRa technology and TES using a system dynamics approach. System dynamics analyzes complex systems by modeling their structure, feedback loops, and time-dependent behaviors. Here, "dynamics" refers to interactions and feedback mechanisms between components that drive system behavior over time, expressed through reinforcing loops (amplifying changes), and balancing loops (stabilizing the system). By analyzing causal relationships and feedback mechanisms, we explore how communication technology choices impact TES performance across reliability, security, and scalability [19]. These insights can inform more effective TES implementations and guide future research.

Our contributions are as follows. First, we present a comprehensive system dynamics analysis of LoRa integration in TES, highlighting key feedback mechanisms that influence system performance. Second, we identify and characterize eleven reinforcing and five balancing loops that shape the technical, economic, and environmental outcomes of LoRa-enabled TES. Third, we introduce a stakeholder integration framework that maps interactions among technology providers, energy producers, consumers, and regulatory bodies. Finally, we provide practical insights for system architects and policymakers by identifying key leverage points and implementation challenges. These findings enhance the theoretical understanding of TES communication infrastructure while offering guidance for system design and policy development.

The remainder of this paper is structured as follows. Section II reviews the literature on TES and LoRa technology, identifying key challenges and opportunities. Section III presents the system dynamics methodology. Section IV examines relationships between system components through causal loop diagrams. Section V discusses the implications of our findings, and Section VI concludes with recommendations for future research.

II. Literature Review

The traditional power grid is undergoing a fundamental transformation, shifting away from its historical one-way electricity delivery from centralized plants to end consumers. This change is driven by environmental concerns (with energy consumption being a predominant source of climate change, accounting for more than 60% of total global greenhouse gas emissions), technological advancements (through introduction of renewable energy sources and cleaner energy technologies), and evolving consumer needs (particularly the need for reliable and affordable energy for basic needs and well-being) [1] [2]. A major catalyst in this transition is the integration of DERs, such as rooftop solar panels, small wind turbines, and battery storage systems, which are positioned closer to consumption points and enable localized energy production [5]. However, integrating DERs into the grid requires a market framework that allows dynamic energy exchange, leading to the emergence of TES.

TES facilitate decentralized electricity trading through automated market mechanisms, empowering prosumers—entities that both consume and produce energy—to optimize their energy usage and trade excess generation at competitive prices [6] [7] [20] [21]. Unlike traditional grid structures, where excess energy must be sold to the main grid under regulated tariffs, TES allows prosumers to engage directly with local consumers, promoting flexibility and efficiency [22]. The effectiveness of TES hinges on several interconnected components: microgrids, which can operate independently or in conjunction with the main grid; smart meters, which provide real-time monitoring and automated trading capabilities; and energy management systems, which optimize energy flow and market transactions [10]. Despite these advantages, TES implementation presents significant challenges, particularly in communication infrastructure, data security, privacy, system scalability, and seamless integration with existing grid operations [12] [13] [18].

The communication backbone of TES plays a pivotal role in enabling secure and efficient transactions. Various technologies have been explored to facilitate information exchange among market participants, including WiFi and cellular networks. While WiFi provides a widely available and cost-effective solution, its limitations—such as range restrictions, high power consumption, and susceptibility to interference—hinder its scalability, particularly in dense urban environments [23]. Cellular networks, on the other hand, offer extensive coverage and reliable data transfer but come with high operational costs and significant energy demands, making them less ideal for large-scale TES deployments [24] [25]. These limitations necessitate alternative communication technologies that can balance energy efficiency, cost, and performance.

LoRa technology has emerged as a promising alternative for TES communication networks. Designed for Internet of Things (IoT) applications, LoRa operates on a Low-Power Wide-Area Network protocol, offering long-range connectivity with minimal power consumption and strong interference resistance [14] [15]. Studies have demonstrated its advantages over traditional IoT communication technologies, highlighting superior energy efficiency, scalability, and security features [16] [17] [26]. However, trade-offs exist between reliability and energy efficiency, as efforts to enhance data transmission reliability often lead to increased energy consumption of end devices [27] [28]. Research on LoRa scalability has shown that network performance can be improved by adding gateways or adopting dynamic transmission strategies, yet increased deployment density may lead to interference and reduced coverage probability [29] [30] [17].

The integration of LoRa into TES introduces complex interactions among technical, economic, and social factors, necessitating a holistic analytical approach. Systems thinking provides a framework for understanding these interdependencies, emphasizing feedback mechanisms and causal relationships within the energy ecosystem [19]. Previous studies applying this approach to energy systems have demonstrated its value in identifying unintended consequences and optimizing the performance of emerging technologies [19]. By leveraging systems thinking methodologies, such as causal loop analysis, we can assess how LoRa influences key TES attributes—efficiency, security, scalability, reliability, and cost-effectiveness—while considering the broader implications of communication technology choices. This perspective is crucial for developing strategic implementation plans and mitigating potential challenges in real-world TES deployments.

III. Methodology

A. A Systems Dynamics Approach to Transactive Energy System Analysis

The intricate structure of TES necessitates an analytical framework that captures both direct interactions and emergent system-wide behaviors. Traditional analytical approaches often fall short in accounting for the interdependencies among market participants, communication networks, regulatory frameworks, and technological infrastructures. Systems dynamics offers a robust alternative, enabling a holistic examination of TES by mapping causal relationships, feedback loops, and evolving system states over time [19]. This approach not only facilitates a deeper understanding of TES operations but also provides a structured methodology for assessing the impact of integrating LoRa technology into these systems.

Our analysis unfolds across three progressive stages: defining system boundaries, identifying stakeholder interactions, and constructing causal loop models. Each stage builds upon the previous, culminating in a comprehensive evaluation of how communication infrastructure—specifically

LoRa—shapes TES performance, scalability, and sustainability.

B. Defining System Boundaries

Establishing clear system boundaries is fundamental to understanding the scope and limitations of the analysis. In this study, the TES ecosystem is conceptualized as a hierarchical framework comprising three distinct layers, as illustrated in Figure 1. At the core technical level, the system integrates DERs, microgrids, and communication infrastructure. This technical foundation operates within a broader system context characterized by market structures, regulatory frameworks, and economic incentives. The external environment, which includes global energy policies, climate goals, and market trends, provides the overarching context that influences the entire system.

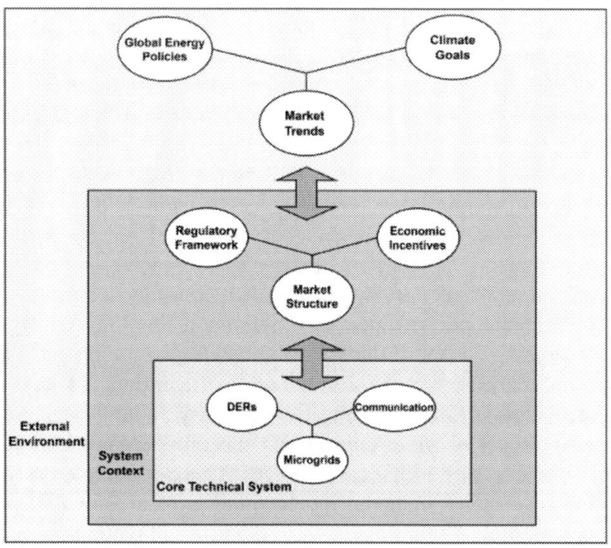

Figure 1. Hierarchical framework of TES: Core Technical System (yellow), System Context (purple), and External Environment (white).

While TES functions within these broader global energy trends, our boundary definition focuses primarily on the interactions between the core technical system and its immediate system context. This approach allows for a targeted investigation into the role of communication technologies while acknowledging the influence of regulatory and market factors that directly impact system operations. The defined boundaries enable us to maintain analytical depth while recognizing the hierarchical nature of TES interactions.

System boundaries are defined based on functional relevance, interdependence, and scalability, but face challenges due to dynamic interactions and evolving technologies. If we were to consider the external environment including global energy policies, market trends and climate goals, our system becomes more complex with multiple variables making it difficult to manage the loops and interactions. For instance, the integration of LoRa as a communication interface is guided by energy efficiency, range, security, and cost-effectiveness. However, boundaries must adapt as TES scale and technology advance, requiring flexible definitions to address emerging interactions and trade-offs. This dynamic approach ensures the analysis remains relevant across different deployment scales and future developments.

The selection of optimal system interfaces, particularly communication means like LoRa, was determined through evaluation of energy consumption profiles, spatial coverage requirements, data throughput needs, and security protocols. One key criterion used for boundary definition was the core technical performance of the LoRa device itself, which helped maintain analytical focus while acknowledging the hierarchical influence of broader factors. The dynamic nature of boundaries becomes particularly important when considering how regulatory changes might trigger cascading effects through market structures into technical operations, or how advancements in communication protocols might enable new market mechanisms that reshape the entire system architecture.

C. Stakeholder Interactions and System Dynamics

The complexity of TES arises from the intricate network of stakeholder relationships that shape system behavior and performance. These interactions, depicted in Figure 2, extend beyond direct transactional exchanges, encompassing regulatory influence, technological dependencies, and sustainability considerations. Understanding these dynamics is essential for identifying intervention points, anticipating systemic responses to change, and fostering resilience in energy markets.

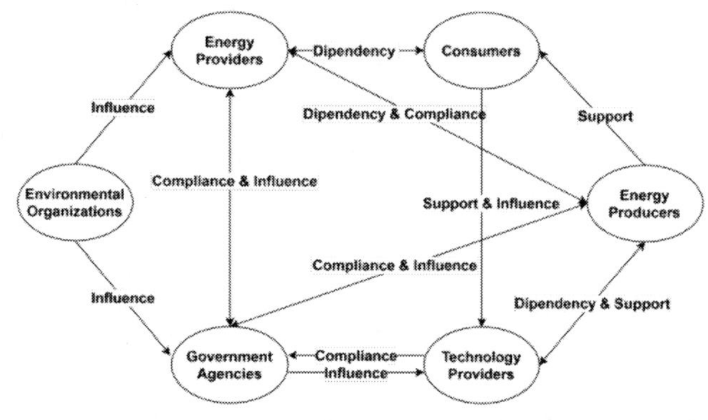

Figure 2. Stakeholder Interest Map in Transactive Energy Systems with LoRa Integration.

The diagram illustrates the complex web of relationships between key stakeholders in TES. Arrows indicate primary interaction types: dependency (resource or service reliance), support (financial or operational assistance), compliance (regulatory or standard adherence), and influence (impact on decision-making or behavior).

At the foundation of TES, the interdependence between Energy Providers and Consumers forms a bidirectional relationship that is fundamental to system stability. Energy Providers ensure a reliable supply of electricity, while Consumers generate the demand that sustains economic viability. This relationship is further shaped by Energy Producers, who serve as both suppliers and market participants. Their operations are driven by demand fluctuations, policy frameworks, and technological advancements that influence production capacity and energy distribution.

Environmental Organizations exert a significant shaping force on the TES landscape, primarily through their advocacy for sustainable energy practices and stringent environmental regulations. Their influence is particularly pronounced in their interactions with Government Agencies and Energy Providers, where they drive policy decisions related to renewable energy integration, emissions reductions, and sustainability reporting. These pressures translate into regulatory mandates that compel stakeholders to align operational strategies with broader environmental objectives.

Government Agencies function as both regulators and facilitators within TES, enforcing compliance while simultaneously shaping market conditions through policy interventions. Their engagement spans multiple stakeholders, including Energy Providers, Producers, and Technology Providers, ensuring that system-wide objectives, such as reliability, equity, and sustainability are maintained. This dual role enables Government Agencies to mediate competing interests, balancing economic viability with regulatory imperatives.

Technology Providers play a crucial role in sustaining and advancing TES, offering infrastructure solutions that facilitate energy transactions, enhance grid reliability, and ensure regulatory compliance. Their relationship with Government Agencies is particularly dynamic, as evolving policy frameworks necessitate continuous technological adaptation. Moreover, their support relationships with Energy Producers underscore the increasing reliance on digitalization and smart grid solutions in energy management.

These interconnected relationships give rise to critical feedback loops that reinforce or modify system dynamics. Notable among these are:

- *The Compliance-Driven Innovation Loop*: Energy Providers, under regulatory pressure from Government Agencies, seek advanced solutions from Technology Providers, leading to innovations that enhance reliability and compliance.
- *The Market Development and Adoption Loop*: Consumer preferences influence Energy Producers, who in turn engage with Technology Providers to adopt solutions that meet emerging market demands, shaping the trajectory of technological advancements.
- *The Sustainability Influence Loop*: Environmental Organizations exert pressure on both Government Agencies and Energy Providers, driving legislative changes that enforce sustainability measures, thereby altering operational and investment priorities.

Recognizing and analyzing these feedback mechanisms is essential for stakeholders aiming to navigate TES complexities effectively. Identifying points of leverage within these interactions can facilitate targeted interventions that enhance system efficiency, promote technological innovation, and support the integration of emerging solutions, such as LoRa technology. Furthermore, a nuanced understanding of stakeholder dynamics enables proactive management of potential resistance to change, ensuring that transitions toward sustainable and resilient energy systems occur smoothly and equitably.

D. Causal Loop Modeling and System Behavior

Building upon stakeholder interactions, we construct causal loop diagrams to map the interdependencies among key system variables, offering a structured perspective on system behavior. This iterative modeling process reveals two fundamental feedback mechanisms: reinforcing loops, which amplify trends within the system, and balancing loops, which introduce constraints that stabilize outcomes. By delineating these relationships, we gain insight into how the integration of LoRa technology influences TES across multiple dimensions.

One crucial aspect is cost dynamics, where energy consumption patterns and infrastructure requirements shape economic feasibility. System performance, particularly communication reliability and security, emerges as another critical factor, influencing the overall resilience of TES. Environmental considerations also come into play, as energy efficiency and resource utilization determine sustainability outcomes. Meanwhile, implementation challenges—including technical requirements and integration complexities—affect the practicality of deploying LoRa technology within existing frameworks.

The resulting causal loop diagram (Figure 3) provides a visual representation of these interactions, highlighting leverage points where targeted interventions can optimize TES performance. By capturing the interconnected nature of TES components, this model facilitates scenario analysis, allowing stakeholders to anticipate system behavior under different implementation strategies.

Adopting a systems dynamics perspective in TES analysis offers several key advantages. It uncovers hidden dependencies that conventional assessments might overlook, bringing potential unintended consequences to light before large-scale deployment. Additionally, it provides a structured framework for evaluating trade-offs between technological capabilities, economic feasibility, and regulatory constraints. More importantly, this approach enables an integrated examination of how communication technology choices shape the long-term evolution of TES, equipping decision-makers with the insights necessary to develop resilient, efficient, and sustainable transactive energy markets.

IV. Systems Analysis Results and Discussion

Our systems dynamics analysis, illustrated in Figure 3, reveals a complex web of interactions between LoRa technology integration and TES performance. Through careful examination of the causal loop diagram, we identify eleven reinforcing mechanisms (R1-R11) and five balancing loops (B1-B5) that shape system behavior across multiple dimensions. These interconnected feedback mechanisms offer deep insights into how communication technology choices influence system evolution.

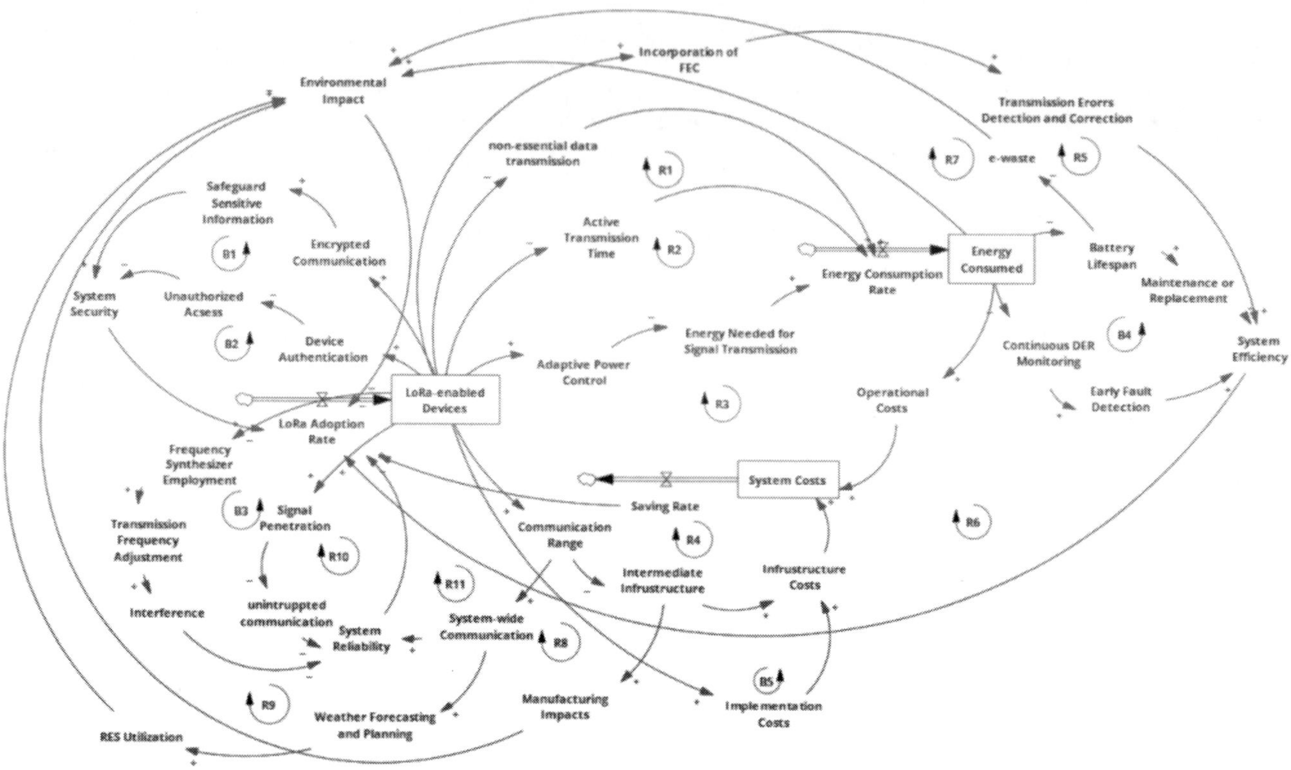

Figure 3. Causal loop diagram of impact of using LoRa on system behavior.

The integration of LoRa technology influences system efficiency through several interconnected pathways. At the core, continuous monitoring of Distributed Energy Resources creates a primary reinforcing loop (R6) that enhances system performance. When energy consumption decreases through LoRa's low-power operation, devices maintain longer monitoring periods without battery replacement. This benefit is amplified by early fault detection capabilities, creating a positive feedback loop that partially offsets the maintenance burden typically associated with large-scale deployments. However, as system scale increases, maintenance requirements (B4) create a counterbalancing effect, suggesting the need for predictive maintenance strategies to optimize this trade-off.

Cost relationships in LoRa-enabled TES demonstrate particularly interesting cascading effects. The technology's long-range capabilities reduce infrastructure requirements through a saving rate mechanism (R4) that interacts with adaptive power control features (R3) to create compound cost benefits. These benefits manifest through reduced operational expenses and decreased infrastructure needs. LoRa also minimizes non-essential data transmission through event-driven communication and scheduled messaging, ensuring data is transmitted only when critical thresholds or specific events occur (e.g., changes in energy demand, supply, or price). This reduces the energy consumption rate (R1) by lowering active transmission time, as LoRa devices operate with low duty cycles, remaining in sleep mode and only waking briefly to transmit or receive data (R2). Adaptive power control further optimizes energy use by dynamically adjusting transmission power and minimizing airtime (R3). Consequently, this cumulative reduction in energy consumption translates into lower operational costs, reinforcing cost savings and enabling further system expansion. However, our analysis reveals a sophisticated balancing mechanism (B5) where implementation costs moderate these advantages through multiple pathways involving infrastructure and manufacturing impacts. This interaction suggests that implementation timing and scaling strategies significantly influences overall cost effectiveness.

The security and reliability aspects of the system reveal previously unexplored connections. LoRa's encrypted communication and device-level authentication create strong initial security benefits (B1, B2), ensuring that sensitive information remains confidential and accessible only to authorized devices. These security mechanisms interact with system adoption dynamics, forming balancing loops (B1, B2), where increased security reduces the marginal benefit of adding new LoRa devices solely for security purposes, eventually slowing down adoption rates. Meanwhile, Chirp Spread Spectrum (CSS) modulation supports uninterrupted communication by enhancing resistance to interference and multipath fading. This ensures strong signal penetration (R10), which remains unaffected even as device density increases, reinforcing system reliability. However, increased device density introduces potential interference, requiring careful network management.

Environmental feedback loops in LoRa-enabled TES demonstrate more nuanced benefits than initially apparent, influencing both sustainability and system efficiency. The reduction in e-waste (R7) connects directly to extended battery lifespan, as longer-lasting batteries reduce replacement frequency, minimizing discarded electronic waste and its environmental impact. System-wide communication improvements (R8) also contribute by reducing the need for intermediate infrastructure, thereby lowering the demand for raw materials and decreasing manufacturing-related emissions. Perhaps most significantly, LoRa's long-range communication enables precise weather forecasting (R9), enhancing renewable energy system utilization. By integrating weather sensors across vast areas, LoRa facilitates real-time, high-resolution data collection, allowing smart grids to optimize energy distribution and storage based on changing environmental conditions. This capability strengthens the reliability of renewable energy sources, accelerating the transition toward a cleaner and more sustainable energy ecosystem. These interconnected mechanisms suggest that environmental benefits may accumulate more rapidly than previously understood, with important implications for long-term sustainability planning.

The multiple pathways affecting system costs and performance indicate the need for careful phasing of technology adoption. Critical intersection points emerge where non-essential data transmission interacts with adaptive power control, suggesting opportunities for optimizing communication strategies. Similarly, the relationship between frequency synthesizer employment and system reliability reveals potential bottlenecks that must be considered in scaling plans.

These interconnected feedback mechanisms suggest several important considerations for LoRa integration in TES. Implementation planning must account for both immediate benefits and longer-term scaling effects, particularly where security measures interact with system reliability. System designers should anticipate and plan for transition points where balancing loops begin to counteract initial benefits, while recognizing that environmental benefits may continue to accumulate through multiple reinforcing loops with fewer balancing constraints.

This systems analysis provides valuable insights for future TES development while highlighting areas requiring further investigation. Particularly important is the need to understand how these feedback mechanisms behave under different deployment scales and environmental conditions, especially considering the complex interactions between security, reliability, and system performance.

V. Limitations and Assumptions

This study's system dynamics analysis is built on several key assumptions and technical constraints that shape its findings. The causal loop analysis presumes relatively stable relationships between system components, allowing for systematic evaluation, though real-world interactions are often more dynamic and nonlinear. It also assumes rational stakeholder behavior, which may not always hold true in complex energy markets. Additionally, the study considers LoRa technology based on its current documented capabilities, though future advancements could shift these parameters, potentially altering the validity of identified feedback mechanisms.

From a technical standpoint, LoRa's limited data rates, ranging from 22 bps to 27 kbps, present significant challenges for real-time data exchange in TES. This constraint becomes particularly problematic during market fluctuations or grid instability, where rapid data transmission is critical. While LoRa's strengths lie in its energy efficiency and long-range capabilities, these benefits come at the cost of substantially lower throughput compared to WiFi or cellular networks. Moreover, LoRaWAN's architecture inherently favors uplink communication from devices to gateways, while downlink capacity remains constrained. This asymmetry complicates TES operations, which require bidirectional communication for control signals, market updates, and system feedback. The combination of low data rates and extended transmission times further limits real-time responsiveness, posing challenges for dynamic price signaling, grid stability management, coordinated demand response, and emergency interventions. Given these constraints, LoRa alone may not fully meet the communication demands of TES, but its advantages in efficiency and coverage suggest that hybrid approaches—integrating LoRa with higher-bandwidth solutions—could offer a more effective path forward.

At a real-world scale, this systems dynamics analysis approach and proposed LoRa integration is applicable in small to medium urban districts, particularly in microgrid networks serving 1,000-5,000 users where phased deployment of TES is planned. The framework is especially suitable for areas requiring low-power, long-range communication coverage with modest bandwidth requirements, and locations with mixed energy generation sources. Smart cities represent another key application area, especially communities developing new energy management systems and cities transitioning to renewable energy sources, where systematic analysis of stakeholder interactions is important and cost-effective communication infrastructure is required.

VI. Conclusion and Future Directions

This systems dynamics analysis of LoRa technology integration in transactive energy systems reveals complex interdependencies that shape system performance and evolution. Through careful examination of feedback mechanisms, we identify several key insights that inform both theoretical understanding and practical implementation of communication infrastructure in modern energy systems.

The analysis demonstrates that communication technology choices influence TES performance through multiple interconnected pathways. LoRa's technical characteristics create both reinforcing and balancing feedback loops across system efficiency, cost, security, and environmental dimensions. While initial benefits in areas, such as energy consumption and infrastructure costs encourage adoption, counterbalancing forces emerge as systems scale. This suggests the existence of optimal implementation scales that maximize benefits while managing complexity.

These findings have important implications for TES development. System architects must consider not only immediate technical benefits but also longer-term scaling effects when designing communication infrastructure. The emergence of balancing loops at larger scales indicates the need for careful planning and potentially hybrid approaches that combine multiple communication technologies to address different system requirements.

Several areas warrant further investigation. First, the causal loop generated in this work still needs rigorous validation. To this end, quantitative modeling could provide more precise understanding of the transition points where balancing loops begin to dominate system behavior. Second, real-world case studies could validate and refine our understanding of these feedback mechanisms under various deployment conditions. Third, investigation of hybrid communication architectures, as recommended in this work, might reveal ways to maintain benefits while mitigating scaling challenges.

Additionally, future research should examine how these system dynamics might evolve under different regulatory frameworks and market structures. The interaction between technical feedback loops and institutional constraints could reveal important considerations for policy development and market design.

The transition to more sustainable and efficient energy systems requires careful consideration of how individual technologies influence overall system behavior. This analysis demonstrates the value of systems thinking in understanding these relationships while highlighting the importance of considering both immediate and longer-term effects of technology choices in complex energy systems.

References

[1] R. G. Newell and D. Raimi, "Global energy outlook comparison methods: 2020 update," *Resources for the Future*, pp. 20–06, 2020.

[2] Y. Sugiawan and S. Managi, "New evidence of energy-growth nexus from inclusive wealth," *Renewable and Sustainable Energy Reviews*, vol. 103, pp. 40–48, 2019.

[3] N. Khraief, M. Shahbaz, H. Mallick, and N. Loganathan, "Estimation of electricity demand function for algeria: Revisit of time series analysis," *Renewable and Sustainable Energy Reviews*, vol. 82, pp. 4221–4234, 2018.

[4] P. Koukaras *et al.*, "Integrating blockchain in smart grids for enhanced demand response: Challenges, strategies, and future directions," *Energies*, vol. 17, no. 5, pp. 1–18, 2024.

[5] M. A. Tabar, M. A. Jirdehi, and M. Shaterabadi, "Impact of bi-facial pv panels' presence as the novel option on the energy management and scheduling of the interconnected grids: Comprehensive outlook," *Journal of Building Engineering*, vol. 90, p. 109495, 2024.

[6] F. Lezama *et al.*, "Local energy markets: Paving the path toward fully transactive energy systems," *IEEE Transactions on Power Systems*, vol. 34, no. 5, pp. 4081–4088, 2019.

[7] M. Daneshvar, M. Pesaran, and B. Mohammadi-Ivatloo, "Transactive energy in future smart homes," in *The Energy Internet*, W. Su and A. Q. Huang, Eds. Woodhead Publishing, 2019, pp. 153–179.

[8] Z. Liu, L. Wang, and L. Ma, "A transactive energy framework for coordinated energy management of networked microgrids with distributionally robust optimization," *IEEE Transactions on Power Systems*, vol. 35, no. 1, pp. 395–404, 2020.

[9] B. H. Rao and P. S. M, "Prosumer participation in a transactive energy marketplace: A game-theoretic approach," in *2020 IEEE International Power and Renewable Energy Conference*, Oct. 2020, pp. 1–6.

[10] S. Zhang, D. May, M. Gül, and P. Musilek, "Reinforcement learning-driven local transactive energy market for distributed energy resources," *Energy and AI*, vol. 8, 2022.

[11] D. Xu *et al.*, "Peer-to-peer multienergy and communication resource trading for interconnected microgrids," *IEEE Transactions on Industrial Informatics*, vol. 17, no. 4, pp. 2522–2533, 2021.

[12] P. Siano, G. Marco, A. Rolán, and V. Loia, "A survey and evaluation of the potentials of distributed ledger technology for peer-to-peer transactive energy exchanges in local energy markets," *IEEE Systems Journal*, vol. 13, no. 3, pp. 3454–3466, 2019.

[13] O. Abrishambaf, F. Lezama, P. Faria, and Z. Vale, "Towards transactive energy systems: An analysis on current trends," *Energy Strategy Reviews*, vol. 26, p. 100418, 2019.

[14] Z. Sun *et al.*, "Recent advances in lora: A comprehensive survey," *ACM Transactions on Sensor Networks*, vol. 18, no. 4, 2022.

[15] J. P. S. Sundaram, W. Du, and Z. Zhao, "A survey on lora networking: Research problems, current solutions, and open issues," *IEEE Communications Surveys and Tutorials*, vol. 22, no. 1, pp. 371–388, 2020.

[16] C. Li and Z. Cao, "Lora networking techniques for large-scale and long-term iot: A down-to-top survey," *ACM Computing Surveys*, vol. 55, no. 3, 2022.

[17] O. Georgiou and U. Raza, "Low power wide area network analysis: Can lora scale?" *IEEE Wireless Communications Letters*, vol. 6, no. 2, pp. 162–165, 2017.

[18] O. Tooki and O. M. Popoola, "A comprehensive review on recent advances in transactive energy system: Concepts, models, metrics, technologies, challenges, policies and future," *Renewable Energy Focus*, vol. 50, 2024.

[19] N. Dhirasasna and O. Sahin, "A system dynamics model for renewable energy technology adoption of the hotel sector," *Renewable Energy*, vol. 163, pp. 1994–2007, 2021.

[20] M. Mallaki, M. S. Naderi, M. Abedi, S. D. Manshadi, and G. B. Gharehpetian, "A novel energy-reliability market framework for participation of microgrids in transactive energy system," *International Journal of Electrical Power and Energy Systems*, vol. 122, 2020.

[21] K. Polat and A. Ozdemir, "Impacts of transactive energy trading on the load point reliability indices of lv distribution system," *IEEE Access*, vol. 11, pp. 132 119–132 130, 2023.

[22] M. Jalali, K. Zare, and S. Tohidi, "Designing a transactive framework for future distribution systems," *IEEE Systems Journal*, vol. 15, no. 3, pp. 4221–4229, 2020.

[23] H. Zhou, B. Li, X. Zong, and D. Chen, "Transactive energy system: Concept, configuration, and mechanism," *Frontiers in Energy Research*, vol. 10, pp. 1–14, 2023.

[24] X. Ge *et al.*, "Spatial spectrum and energy efficiency of random cellular networks," *IEEE Transactions on Communications*, vol. 63, no. 3, pp. 1019–1030, 2015.

[25] L. Fu *et al.*, "Joint optimization of multicast energy in delay-constrained mobile wireless networks," *IEEE/ACM Transactions on Networking*, vol. 26, no. 1, pp. 633–646, 2018.

[26] S. Devalal and A. Karthikeyan, "Lora technology—an overview," in *2020 International Conference on Communication and Signal Processing*, July 2020, pp. 1215–1219.

[27] X.-C. Le, B. Vrigneau, M. Gautier, M. Mabon, and O. Berder, "Energy/reliability trade-off of lora communications over fading channels," *IEEE Communications Letters*, vol. 25, no. 3, pp. 1015–1019, 2021.

[28] S. S. Borkotoky, J. F. Schmidt, U. Schilcher, P. Battula, and S. Rathi, "Reliability and energy consumption of lora with bidirectional traffic," *IEEE Communications Letters*, vol. 25, no. 11, pp. 3743–3747, 2021.

[29] M. Bor, U. Roedig, T. Voigt, and J. M. Alonso, "Do lora low-power wide-area networks scale?" in *MSWiM 2016 - Proceedings of the 19th ACM International Conference on Modeling, Analysis and Simulation of Wireless and Mobile Systems*, Nov. 2016, pp. 59–67.

[30] R. Marini, K. Mikhaylov, G. Pasolini, and C. Buratti, "Lorawansim: A flexible simulator for lorawan networks," *Sensors*, vol. 21, no. 3, pp. 1–19, 2021.

International Academy, Research, and Industry Association (IARIA)
412 Derby Way
Wilmington, DE 19810

ISBN 979-8-3313-1562-7

CURRAN ASSOCIATES INC.
proceedings.com